NUREG/CR-7145

Nuclear Power Plant Security Assessment Guide

Office of Nuclear Security and Incident Response

AVAILABILITY OF REFERENCE MATERIALS
IN NRC PUBLICATIONS

NUREG/CR-7145

Nuclear Power Plant Security Assessment Guide

Manuscript Completed: April 2013
Date Published: April 2013

Prepared by:
J. Zamanali and C. Chwasz

Nuclear Systems Analysis Operations Center
Information Systems Laboratories, Inc.
Rockville, MD 20852

James E. Vaughn, NRC Project Manager

NRC Job Codes N4110 and N4116

Office of Nuclear Security and Incident Response

ABSTRACT

This document provides detailed guidance for the format and content of a security assessment of a commercial nuclear power plant.

The U.S. Nuclear Regulatory Commission (NRC) encourages design certification and combined license applicants to use this guidance to optimize physical security during the design phase. The expected result is a more robust security posture with less reliance on operational programs (human actions) and potentially costly retrofits. The NRC also encourages operating reactor licensees to use this guidance in planning and executing changes and upgrades of physical protection systems at existing sites.

CONTENTS

LIST OF FIGURES

LIST OF TABLES

EXECUTIVE SUMMARY

In a staff requirements memorandum (SRM) to SECY-06-0204, dated April 24, 2007 (Ref. 6), the U.S. Nuclear Regulatory Commission (NRC) staff was directed by the Commission to develop a security assessment guidance document. The Nuclear Power Plant Security Assessment Format and Content Guide was finalized and entered into the agency's Agencywide Documents Access and Management System (ADAMS) document database on October 30, 2007. New reactor designers used this document in the preparation of applications for design certifications. In September 2011, the NRC Office of Nuclear Security and Incident Response, Security Programs Support Branch, initiated an action to convert the Nuclear Power Plant Security Assessment Format and Content Guide into an official NRC NUREG/CR.

The NUREG/CR expanded upon the existing security assessment guidance document and was informed by lessons learned and requests for additional information (RAIs) generated during the NRC staff review of security assessments submitted by design certification and combined license applications between 2006 and 2012.

This document provides detailed guidance for the format and content of a security assessment. The security assessment described in this document is voluntary for licensees and applicants for nuclear power plants. This document should be used in conjunction with the Nuclear Power Plant Security Assessment Technical Manual, SAND2007-5591, September 2007 (Ref. 7). The Standard Review Plan, NUREG-0800, "Standard Review Plan for the Review of Safety Analysis Reports for Nuclear Power Plants (LWR Edition)" (Ref. 14), Sections 13.6.1, and 13.6.2, also may be used as an aid when performing the security assessment.

The NRC encourages design certification and combined license applicants to use this guidance to optimize physical security during the design phase. The expected result is a more robust security posture with less reliance on operational programs (human actions). Although not required under 10 CFR 52.79, "Contents of Applications; Technical Information in Final Safety Analysis Report," as part of the application for a new nuclear power plant, the security assessment that would result from the method described within this document would provide a strong basis for meeting the general performance objective in 10 CFR 73.55(b)(1). The NRC also encourages operating reactor licensees to use this guidance in planning and executing changes and upgrades of physical protection systems at existing sites.

ACKNOWLEDGMENTS

The authors would like to acknowledge the NRC staff members who provided direction, suggestions, and other assistance in the preparation of this publication.

Office of Nuclear Security and Incident Response

- James E. Vaughn, Security Specialist, Security Programs and Support Branch

- John G. Frost, Security Specialist, Reactor Security and Licensing Branch

- Al Tardiff, Sr. Security Specialist, Fuel Cycles and Transportation Security Branch

- Doug Huyck, Branch Chief, Security Programs and Support Branch

- Dyrk Greenhalgh, Vulnerability Assessment Team Lead, Oak Ridge National Laboratory

- Jack Crockett, Physical Security Engineer, Oak Ridge National Laboratory

ACRONYMS AND ABBREVIATIONS

ADAMS	Agencywide Documents Access and Management System
ANSI	American National Standards Institute
ASSESS	Analytic System and Software for Evaluating Safeguards and Security
ATLAS	Adversary Time Line Analysis System
BREs	bullet resistant enclosures
CAS	central alarm station
CDP	critical detection point
CFR	*Code of Federal Regulations*
CIP	Critical Interruption Point
COL	combined license
DBT	design-basis threat
DC	design certification
DCD	Design Control Document
EASI	Estimated Adversary Sequence Interruption
FOF	force-on-force
JCATS	Joint Conflict and Tactical Simulation
LOCA	loss of coolant accident
MOX	mixed oxide
NEI	Nuclear Energy Institute
NRC	U.S. Nuclear Regulatory Commission
OCA	owner-controlled area
PA	protected area
P_D	probability of detection
P_I	probability of interruption
P_N	probability of neutralization
PPS	physical protection system
PRA	probabilistic risk assessment
RAI	request for additional information
RG	Regulatory Guide
RIS	Regulatory Issue Summary
ROWS	remotely operated weapon system
RSD	required standoff distance
SAS	secondary alarm station
SECY	U.S. NRC Office of the Secretary
SGI	Safeguards Information
SNM	special nuclear material
SRM	staff requirements memorandum
SSCs	structures, systems, and components
VA	vulnerability analysis
VBS	vehicle barrier system
VISA	Vulnerability Integrated Security Assessment

1. INTRODUCTION

1.1 Purpose and Applicability

This guide describes a method for developing the format and content of a security assessment for a nuclear plant. While this document has been developed specifically for new nuclear power plant design certification and combined license applicants, the high assurance evaluation described may be applied by existing nuclear power plant licensees when upgrading or modifying their physical protection systems.

The security assessment is an examination of security in a holistic manner, considering the facility design (including the layout of the facility) and physical characteristics of the site. It may serve as part of the technical bases for evaluating the applicant's security program during the licensing phase. Specifically, the security assessment is an evaluation of the reactor facility's physical protection design that: 1) identifies target sets and, for selected scenarios, performs a systematic evaluation using risk evaluation methodologies that demonstrate the ability of the design to meet the performance objectives of 10 CFR 73.55(a) (Ref. 1), and 2) identifies engineered security design features to be incorporated into the design of the reactor facility that provide high assurance that security functions can be accomplished, to the maximum extent practical, without undue reliance upon administrative and operational response actions by response forces.

The primary purpose of the security assessment is to demonstrate that the physical protection system (PPS) design of a new reactor facility provides high assurance of protection against the design-basis threat (DBT). The performance-based physical security requirements to protect against the DBT can be found in 10 CFR 73.55. Performance of a security assessment does not obviate the requirements of 10 CFR Part 73, "Physical Protection of Plants and Materials" (Ref. 2), for force-on-force (FOF) performance assessments for operating nuclear power plants. The FOF performance assessment serves as a validation tool for the security assessment in that the security assessment should form the basis for the physical protection strategy and identification of target sets. The standard set of DBT scenarios (Ref. 3) provided by the U.S. Nuclear Regulatory Commission (NRC) encompasses the attributes and characteristics of the adversaries as defined in the DBT for radiological sabotage.

If an applicant submits a design-specific security assessment as part of a design certification license or combined license application, the NRC staff will review it to ensure that the design features identified and described are consistent with the relevant security requirements and that practicable safety and security features have been appropriately considered for integration into the design for new reactors (consistent with the Commission Policy Statement on Regulation of Advanced Reactors – 73 FR 60612; October 14, 2008) (Ref. 4).

Resolution of security-related design issues at the early stage of the regulatory review process should result in a more robust security posture requiring less reliance on human actions. However, resolution of the security-related design issues would not constitute final NRC approval of an applicant's overall security program. NRC review and approval of an applicant's security program consists of licensing reviews of an application under 10 CFR Part 50 or 10 CFR Part 52, to include security plans, implementation schedules, and additional security technical reports. Inspections during construction and startup, as well as the security baseline inspection program (to include FOF exercises) provide assurance of compliance with all applicable regulations, orders, and licenses.

1.2   Background

Following the terrorist attacks on September 11, 2001, the NRC conducted a thorough review of security to ensure that nuclear power plants and other licensed facilities continued to have effective security measures in place given the changing threat environment. Through a series of orders, the Commission specified additional supplementary information to be included in the current DBT, as well as requirements for training enhancements, access authorization enhancements, restrictions on security officer work hours and enhancements to defensive strategies, mitigative measures, and integrated response. Since then, the NRC has assessed threats, vulnerabilities, and mitigative strategies for reactor facilities and has required upgrades of physical security and mitigative measures at operating reactors.

The Commission also directed NRC staff to make conforming changes to other rules. As part of this effort, the NRC staff examined implementation issues such as the need for guidance documents, changes in the enforcement policy, and a means for dispositioning the post-September 11, 2001 orders. Furthermore, the Commission directed the NRC staff to provide guidance so that applicants and prospective applicants would be able to use DBT information early in the design stage of new reactor facilities to identify potential mitigative measures and design features. In the staff requirements memorandum (SRM) dated September 9, 2005 titled "Security Design Expectation for New Reactor Licensing Activities," (Ref. 5), the Commission stated that applicants should be required to submit a security assessment addressing the relevant security requirements, which were established for currently operating plants by order, including the requirements for protection against the revised DBT and the requirements for enhanced mitigative measures.

In the SRM dated April 24, 2007 titled "Proposed Rulemaking — Security Assessment Requirements for New Nuclear Power Reactor Designs," (Ref. 6), the Commission directed the NRC staff to develop the guidance for the proposed rule to support the 10 CFR 73.55 power reactor security rulemaking ongoing at the time, and to provide regulatory guidance for licensees to meet the requirements of that rulemaking. This document, in conjunction with the "Nuclear Power Plant Security Assessment Technical Manual" (Ref. 7), represents the guidance developed for security assessments. Included in this document are the insights from the Design Certification, Combined License and ongoing operating license reviews conducted between 2007 and 2012.

1.3   Use of Standard Format and Content

This document describes a process to ensure that the PPS of the reactor facility design is effective in protecting against the DBT with high assurance.

This guide provides an acceptable means of providing the information for nuclear power reactor security assessment submittals, and a uniform format that the NRC staff considers acceptable for structuring and presenting the required information.

The level of detail in the security assessment should be sufficient to enable the NRC staff to understand and determine the validity of all input data and calculation models used, to enable the NRC to understand the sensitivity of the results to key aspects of the PPS including key analysis assumptions (e.g., identification of critical assumptions for which small changes could significantly affect the overall effectiveness of the PPS), and to audit the calculations. The design information provided in the security assessment should reflect the most advanced state of the design at the time of submission. It is not necessary to submit all documentation for NRC review, but basis documents, calculations, guidance, and references should be cited and should be available in a clear, methodical, and retrievable format. Properly retained documentation should allow an independent expert analyst to reproduce any portion of the results or calculations in a straightforward, unambiguous manner. To the extent possible, the retained documentation should be organized along the lines identified in the areas of review.

1.4   Quality Assurance and Security Assessment Team Attributes

1.4.1   Quality Assurance Recommendations

The security assessment should be complete, defensible, and transparent (i.e., traceable to its source documents). Table 1-1 lists the applicable quality assurance recommendations for this analysis. These quality assurance topics are consistent with the quality assurance criteria contained in Appendix B, "Quality Assurance Criteria for Nuclear Power Plants and Fuel Reprocessing Plants," to 10 CFR Part 50 (Ref. 8).

Table 1-1 Quality Assurance Recommendations

	Topic	Recommendation
1	Quality Assurance Program	At the earliest practicable time, consistent with the schedule for developing, modifying, and maintaining the security assessment, a quality assurance program should be established with written policies, procedures, or instructions and should be carried out throughout the life cycle of the assessment.
2	Analysis Staff	Measures are established to provide for indoctrination, training, and qualification of personnel performing security assessment-related activities to assure awareness in quality assurance processes and controls and to ensure suitable technical proficiency is achieved and maintained.
3	Independent Reviews	The security assessment control measures should provide for verifying or checking the adequacy of the assessment, such as by the performance of independent checks and peer reviews. The independent verification or checking process should be performed by individuals or groups other than those who performed the original assessment, but may be from the same organization.
4	Procedures	Activities affecting security assessment quality are prescribed by documented instructions or procedures and should be accomplished in accordance with these instructions or procedures.
5	Document Control	Measures are established to control the issuance of security assessment documents. These measures should ensure that documents, including changes, are reviewed for adequacy and approved for release by authorized personnel. Changes to documents are reviewed and approved by the same organizations that performed the original review and approval unless assigned to another responsible organization.

Table 1-1 Quality Assurance Recommendations

	Topic	Recommendation
6	Corrective Actions	Measures are established to ensure that conditions adverse to security assessment quality are promptly identified and corrected. In the case of significant conditions adverse to quality, the measures should ensure that the cause of the condition is determined and corrective action is taken to preclude repetition. The identification of the significant condition adverse to quality, the cause of the condition, and the corrective action taken is documented and reported to appropriate levels of management.
7	Audits	A comprehensive system of planned and periodic audits is carried out to verify compliance with all aspects of the quality assurance program and to determine the effectiveness of the program. The audits are performed in accordance with written procedures or checklists by appropriately trained personnel not having direct responsibilities in the areas being audited. Audit results should be documented and reviewed by management having responsibility in the area audited. Follow-up action, including re-audit of deficient areas, should be taken where indicated.

1.4.2 Security Assessment Team Recommendations

Team qualification is an important element supporting the credibility and adequacy of the security assessment. Each team member should have technical expertise in the elements that he or she develops. Therefore, the security assessment team should have subject matter experts knowledgeable in the following areas:

1.   Security Systems, to include, but not limited to:
      a.   Detection and Assessment
      b.   Alarm Communication and Display
      c.   Sensors
      d.   Access Control
      e.   Delay
      f.   Communication Systems
      g.   Cyber Security

2.   Protective Force (i.e., Responsive Force)

3.   Others as needed to perform specific functions

Subject matter experts that should be used for the creation and maintenance of target sets are described in RG 5.81, "Target Set Identification and Development for Nuclear Power Reactors" (Ref. 9), section C.3.

The team members, their qualifications, and their roles should be included in the security assessment submittal and supporting documentation.

2. HIGH ASSURANCE EVALUATION GUIDANCE

This section provides a description of an assessment method to evaluate the effectiveness of the designed physical protection system (PPS). This evaluation should demonstrate that the reactor facility's physical protection system elements provide defense-in-depth through the integration of systems, technologies, programs, equipment, supporting processes, and implementing procedures to ensure that the capabilities to detect, assess, interdict, and neutralize[1] threats up to and including the design-basis threat of radiological sabotage are maintained at all times (consistent with the general performance requirements in 10 CFR 73.55(b)(3)). These requirements are considered to meet the objective of high assurance (as defined in 10 CFR 73.55(b)), such that the reactor facility, including the PPS elements, provides high assurance of protection against the DBT.

2.1 Overall Description of the High Assurance Evaluation, Security Assessment Process

The overall security assessment process is depicted in Figure 2-1 and explained further in this chapter by outlining the objectives, process, and reporting recommendations and acceptance criteria of the security assessment process (Ref. 10).

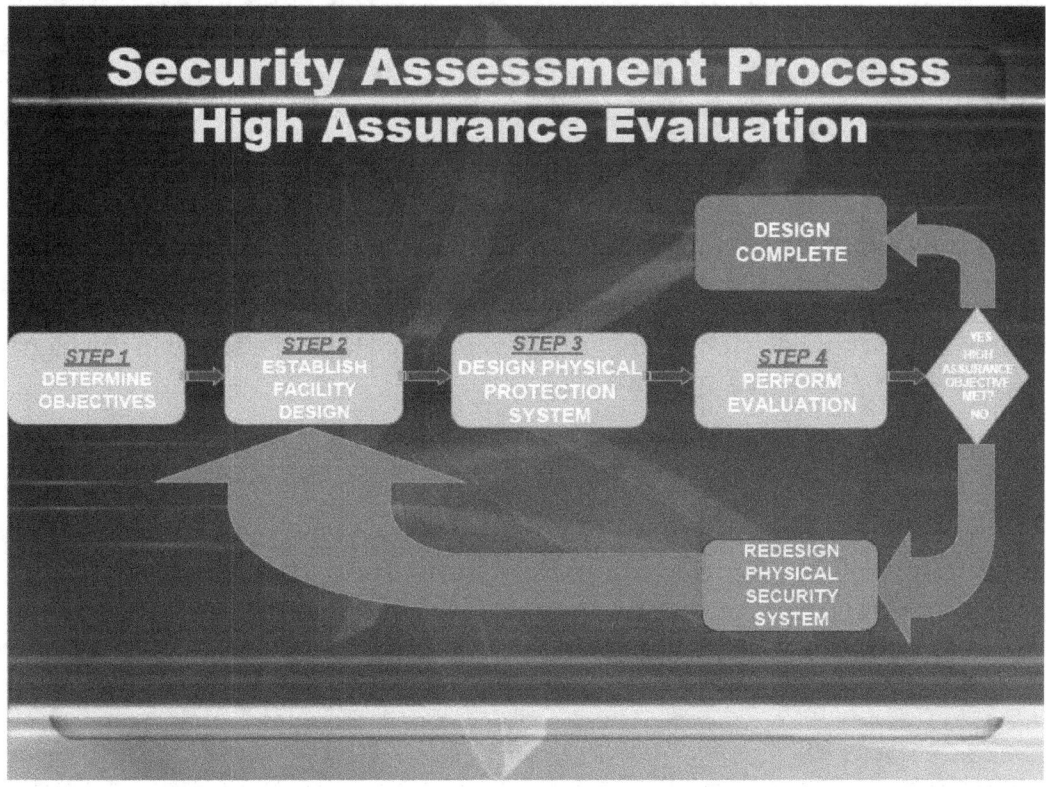

Figure 2-1 Security assessment process – high assurance evaluation

---

[1] For the purposes of this document, the phrase "detect, delay and respond" are used to reflect the process used in security assessments and security assessment software. This phrase is synonymous with the requirement in 10 CFR 73.55(b)(3)(i) for "detect, assess, interdict, and neutralize".

The security assessment process is an iterative effort for the applicant. The NRC, through regulation and guidance documents, identifies the protection requirements and performance standards (Step 1), while the the applicant or licensee determines the protective strategy to accommodate site-specific configuration and operations (Step 2), as well as identify targets, or assets. Step 3 addresses the additional PPS characteristics that were not a part of the existing facility design and adds these to the design such that the combination of reactor facility design and these PPS elements would comply with 10 CFR 73.55, "Requirements for Physical Protection of Licensed Activities in Nuclear Power Reactors Against Radiological Sabotage." The PPS elements, along with the previously established facility and site attributes (Step 2), will be evaluated (Step 4) using a security assessment tool or method of assessment acceptable for use. The evaluation will test whether the PPS elements meet the objective of high assurance (Step 5). If the objective of high assurance has been met, the reactor facility including PPS design should be considered to be in compliance with the 10 CFR 73.55 PPS requirements, and the security design characteristics of the plant are acceptable for the purposes of the security assessment[2]. Otherwise, the reactor facility or PPS design elements of the facility are modified, and the security assessment process is applied again (Step 6).

Changes may be made to the reactor facility design characteristics (Step 2) and to the PPS elements (Step 3). This revised combination of design features, including the PPS elements, would again be evaluated (Step 4) against the objective of high assurance. This iterative process should continue until the reactor facility design meets the objective of high assurance. The overall goal of the iterative security assessment process is to efficiently and effectively achieve a reactor facility design, including the PPS elements that meet the objective of high assurance. Furthermore, as designs are developed, another goal is to provide physical protection through design features, rather than operational functions, thereby reducing dependence on security operational programs (i.e., human actions).

2.2   Determine Objectives

This section describes the security engineering documentation DBT and guidance, and standard set of DBT scenarios (Ref. 3) that are provided by NRC. This step in the high assurance evaluation is depicted in Figure 2-2.

---

[2] Acceptability of the facility and PPS design for 10 CFR 73.55 requirements for NRC licensing.

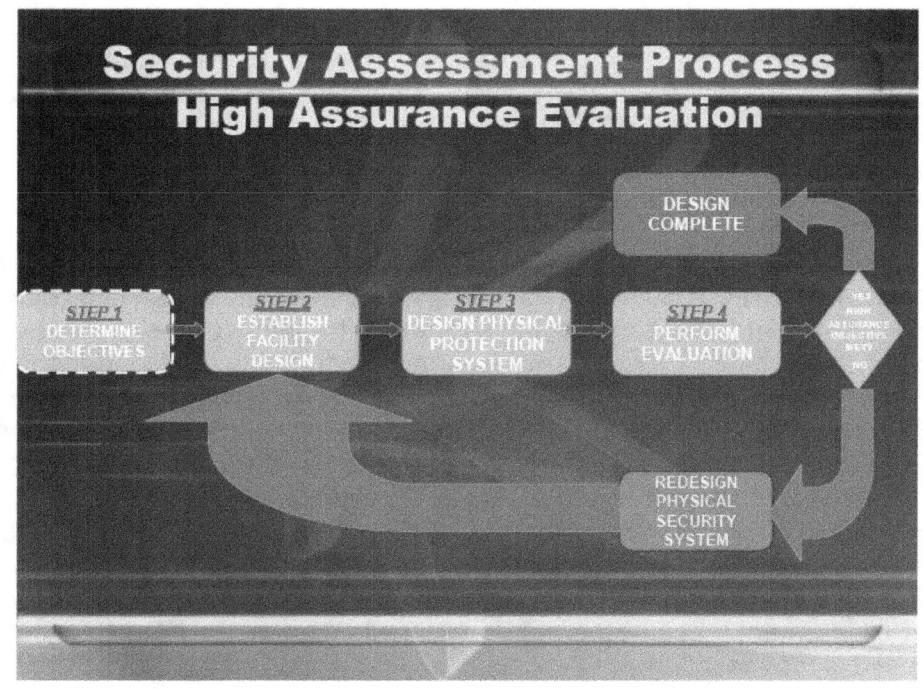

Step 1: Determine Objectives

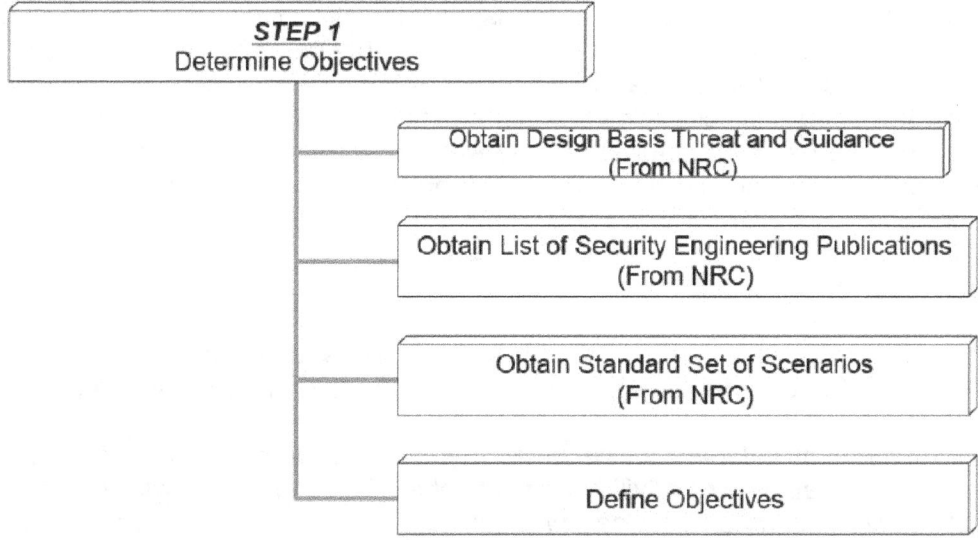

Figure 2-2 Security assessment process – step 1: determine objectives

2.2.1   Obtain Design-Basis Threat and Guidance

Obtain the described radiological sabotage DBT and DBT guidance established by the NRC. Defining the threat establishes the performance that is required from the PPS (Ref. 10). By describing the threat, the assumptions that are made to perform the assessment are documented and are used to show how they influence required upgrades. The DBT is a general description of the threat, including the type of adversaries, the tactics associated with the threat, and the "tools" the adversary may use. Guidance for the DBT includes

10 CFR 73.1(a)(1) (Ref. 11), Radiological Sabotage and Regulatory Guide (RG) 5.69, "Guidance for Application of the Radiological Sabotage Design Basis Threat in the Design, Development, and Implementation of a Physical Security Protection Program that Meets 10 CFR 73.55 Requirements" (Ref. 12).

2.2.2 Obtain List of Security Engineering Publications

Licensees and applicants are encouraged to use the current set of NRC identified security engineering publications when assessing the effectiveness of the designed physical protection system. These publications provide acceptable methods for many of the variable inputs needed during the security assessment process. For example, documents related to the mitigation of vehicles used by an adversary, such as NUREG/CR-6190, "Protection Against Malevolent Use of Vehicles at Nuclear Power Plants" (Ref. 13), detail required (or minimum) standoff distances and vehicle barrier system (VBS) requirements that can support the high assurance evaluation. If applicants or licensees choose to use different publications or resources than those in the reference list, a detailed justification and an explanation of the other publications' adequacy should be provided. Security engineering publications from the NRC that have been identified as acceptable for use in the security assessment process are included as Appendix B. Before conducting the assessment process, the applicant should contact the NRC to ensure the reference list used is the most current.

2.2.3 Obtain Standard Set of Scenarios

Obtain the current standard set of DBT scenarios from the NRC (Ref. 3). These scenarios include specific attributes and characteristics of the DBT that can be used by all applicants for consistency and to ensure that an adequate breadth of the DBT is evaluated. Typical characteristics described will include the number of adversaries, the type of weapons and tools used, the number of teams, and number of adversary entry points. A scenario provides the information necessary to evaluate the performance of the detection, delay, and response elements of the PPS (Ref. 14).

2.2.4 Define Objectives

Before starting a security assessment, it is critical to understand protection system objectives. In 10 CFR 73.55(b), *General performance objective and requirements*, it states that:

(1)   *The licensee shall establish and maintain a physical protection program, to include a security organization, which will have as its objective to provide high assurance that activities involving special nuclear material are not inimical to the common defense and security and do not constitute an unreasonable risk to the public health and safety.*

(2)   *To satisfy the general performance objective of paragraph (b)(1) of this section, the physical protection program must protect against the design basis threat of radiological sabotage as stated in § 73.1.*

(3)   *The physical protection program must be designed to prevent significant core damage and spent fuel sabotage.*

In addition, those applicants anticipating the utilization of mixed oxide (MOX) fuel assemblies will also have as an objective the prevention of theft or diversion of un-irradiated MOX fuel assemblies (Ref. 15).

The objective of high assurance is met when the licensee's protective strategy provides defense-in-depth through the integration of systems, technologies, programs, equipment, supporting processes, and implementing procedures to ensure that the capabilities to detect, assess, interdict, and neutralize threats up to and including the design-basis threat of radiological sabotage are maintained at all times.[3] Detailed guidance for the development of target sets is found in RG 5.81 (Ref. 9).

High assurance can be simulated through the calculation of an overall system effectiveness of the PPS. The overall system effectiveness is a probabilistic calculation of the effectiveness of the PPS to detect (and assess) the adversary, delay the adversary such that responders can intercept the adversary ideally from their protective positions, and the probability that the responders neutralize the adversary. Scenarios in which the responders reach their required protective positions with adequate margin, such that they prevent the adversary from disabling one or more targets, will likely result in a high calculated overall system effectiveness. In comparison, an overall system effectiveness value of 0.0 indicates that the overall PPS is ineffective in stopping the adversary, whereas a 1.0 indicates that the adversary will always be stopped from completing their objectives. Using these overall system effectiveness values, deficiencies and improvements to the physical protection program can be identified and corrected or implemented so that the physical protection system meets the general performance requirements and attains the objective of high assurance. The applicant should clearly state the objective of high assurance and its bases used in the security assessment.

Mathematically, overall system effectiveness can be defined as a probability as follows:

$$P_E = P_I \times P_N \hspace{4cm} \text{Equation (1)}$$

where:

P_E = probability of effectiveness or overall system effectiveness,

P_I = probability of interruption of the adversary (considers the likelihood that detection will occur early enough in the adversary attack sequence that the response force can arrive before the attack is successfully completed), and

P_N = probability of neutralization of the adversary.

Further discussion of the objective of high assurance can be found in Section 2.5.

2.3   Establish Facility Design

Figure 2-3 depicts Step 2 of the high assurance evaluation process. This step establishes the reactor facility design in significant detail to identify potential adversary targets and then identifies these targets using a target set analysis. At this stage, the facility design does not need to include the PPS elements. These elements are added in Step 3 of the assessment process.

---

[3]   "High Assurance," as used in 10 CFR 73.55(a), is deemed to be comparable to the degree of assurance contemplated by the Commission in its safety review for protection against severe postulated accidents having potential consequences similar to the potential consequences from reactor sabotage (44 FR 68185, November 28, 1979).

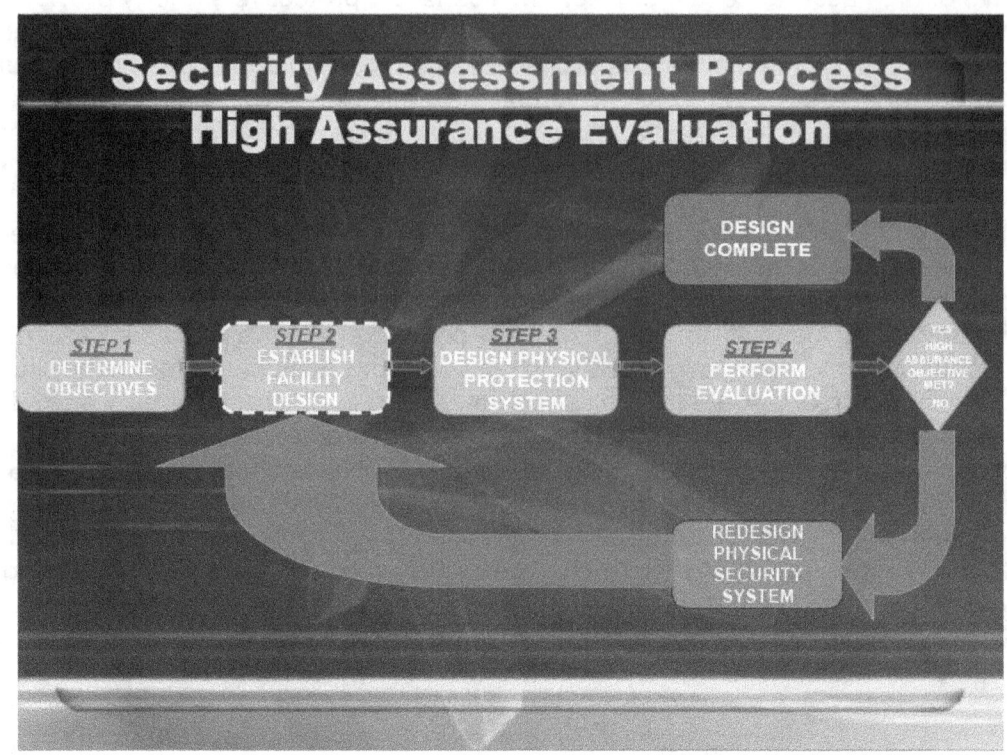

Step 2: Establish Facility Design

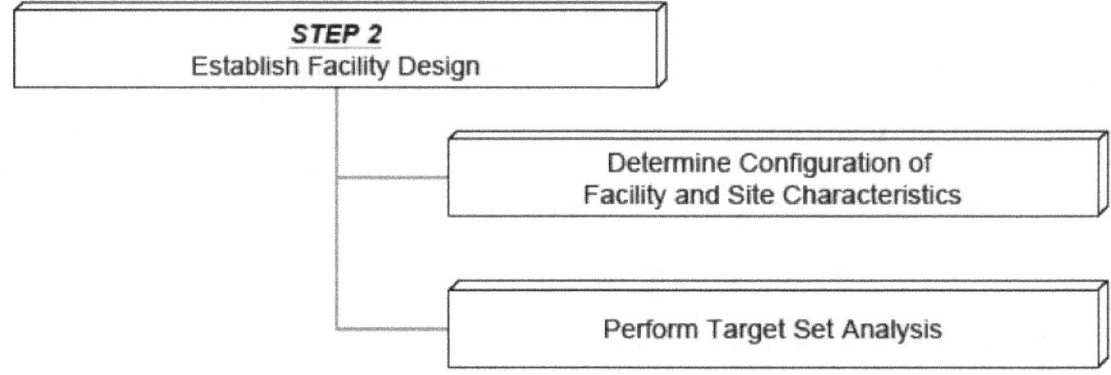

Figure 2-3 Security assessment process – step 2: establish facility design

Applicants for a design certification (DC) should be able to use this process in parallel with other design activities when performing the design process. Applicants for a combined license (COL) referencing a certified design can use this process to identify departures from or additions to the DC. Operating reactor licensees should use this process in parallel with other change processes at the plant (long-term maintenance, upgrades, facility redesign, etc...). Site-specific target sets should be created and used because changes or refinements to the design, based on site characteristics, during the design change process could result in cooresponding changes to identified targets. Therefore, the target set analysis should be reviewed and updated as necessary before COL submission and throughout the review period. During the initial evaluation of the target set analysis and during updates, applicants and licensees are encouraged to optimize the facility design and minimize the reliance on operational security programs, to the maximum extent practical.

The security assessment process should be iterative. As changes are made to the design of the reactor facility, including the PPS elements, physical protection program performance should be evaluated and optimized. Applicants are expected to begin the assessment with an existing facility design as early in the design process as practical. For COL applicants, the facility design to be assessed most likely will be that which has been approved in the Design Control Document. DC applicants, for which approval is pending, most likely will use the submitted design. In all cases, the initial reactor facility design that is being used at the beginning of this iterative process has been termed as the "existing" design throughout this guide.

2.3.1   Determine Configuration of Facility and Site Characteristics

A facility characterization considers the existing nuclear reactor design, including the supporting systems, structures and components; and site characteristics such that the targets at the facility that need to be protected can be identified. The characterization, at this stage, requires limited information about the PPS. The information required would only be that which is required to identify or screen targets. The PPS elements will be added during Step 3 of the process.

The extent of the facility characterization process depends on the design status of the facility. Applicants submitting a DC would only consider the configuration of the design characteristics of the facility. COL applicants and operating reactor licensees will include facility design, site characteristics, and the security operational programs. Note that DC applicants may perform a more comprehensive security assessment by identifying standard physical security characteristics. For example, standard physical security characteristics for a DC applicant could include the location and distance of the protected area perimeter in relation to vital areas. Elements associated with the topography of the terrain or the geographic location of the site or other site- or applicant-specific enhancements or constraints would need to be addressed by COL applicants.

Site characteristics or site parameters that are either postulated in the security assessment or are security design features that are outside the scope of the design being addressed at the particular stage of the regulatory process should be identified as security assessment parameters. These parameters would be addressed by a future applicant that references the design and the assessment. Ultimately, any security design issue identified by an assessment, but not addressed by a security design feature at any application stage, should be identified as a security assessment parameter and should be addressed during the development of the security operational programs under the provisions of 10 CFR Part 73.

The following provides guidance on the scope of the security assessment based on the particular stage of the application process in 10 CFR Parts 50 and 52.

1. Construction Permit (10 CFR Part 50). At the construction permit stage, an applicant would have selected a design and the site on which to build the plant. The scope of the assessment should include a description of the applicant's plan for conducting a security assessment that describes the security design features incorporated into the final design of the site based on the design and site characteristics. Scenarios that necessitate evaluation of the security operational programs would be outside the scope of this assessment. An applicant may choose to postulate location and number of armed responders to perform a more comprehensive security assessment. Any security design issue identified but not addressed by a security design feature, would be recorded as a security assessment parameter and should be addressed by the future operating license applicant.

2. Operating License (10 CFR Part 50). Generally, applicants for a construction permit and an operating license are the same entity. At the operating license stage, the applicant would have developed the security operational programs. The scope of the assessment should include: (1) reference to the security assessment for the construction permit, (2) a description of how security design features left unresolved (security assessment parameters) at the construction permit stage were resolved, and (3) scenarios that necessitate evaluation of the security operational programs. Ultimately, any security design issue identified by the assessment that is not resolved by a security design feature should be identified by a security assessment parameter and must be resolved by the operational security program.

3. Design Certification (10 CFR Part 52). At the design certification stage, the applicant would know the design, but not the site or the security operational programs. The scope of the security assessment should include a description of the applicant's plan for conducting a security assessment and describe the security design features incorporated into the design based on the scenarios evaluated by the assessment. Scenarios that necessitate evaluation of site characteristics and the security operational programs would be outside the scope of this assessment. However, the applicant may decide to assess the effectiveness of the plant's security design features at a hypothetical site or sites having characteristics that fall within a set of postulated site parameters (e.g., the location of transportation routes, heat sink, water access ways, and vehicle pathways). A standard set of physical security characteristics may also be used (e.g., distance from the protected area (PA) barrier, to vital areas with delay and detection, VBS at required (or minimum) standoff distance (as described in Reference 1), and number and location of armed responders). Any security design issue identified but not addressed by a security design feature should be recorded as a security assessment parameter and addressed by a future applicant that references the design certification.

4. Manufacturing License. An applicant for a manufacturing license that references a design certification for which a security assessment was done would know the design but not the site or the security operational programs. However, the manufacturing license applicant would not change the information in the design certification. Therefore, a security assessment would not be required at the manufacturing license stage. Any security design issue identified but not addressed by a security design feature at the design certification stage should continue to be recorded as unresolved and addressed by a future applicant that references the manufacturing license.

If the manufacturing license application proposes to use a custom design (i.e., does not reference a design certification), then the scope of the assessment would be the complete design. Any security design issue identified but not addressed by a security design feature should be recorded as a security assessment parameter and addressed by a future applicant that references the manufacturing license.

5. Standard Design Approval. At the standard design approval stage, the applicant would know the design, but not the site or the security operational programs. The application must include a description of the applicant's plan for conducting a security assessment that describes the security design features incorporated into the design based on the scenarios evaluated by the assessment. Scenarios that necessitate evaluation of site characteristics and the security operational programs would be outside the scope of this assessment. However, the applicant may desire to assess the effectiveness of the plant's security design features at a hypothetical site or sites having characteristics that fall within a set of postulated security assessment parameters (e.g., the location of transportation routes, heat sink, water access ways, and vehicle pathways). Any security design issue identified but not addressed by a security design feature should be recorded as a security assessment parameter and addressed by a future applicant that uses the standard design approval, in developing its operational security program.

6. Combined License (COL) (10 CFR Part 52). An applicant for a COL that selects a plant design by referencing either a design certification or manufacturing license, for which a security assessment was completed, would have knowledge of the design, the site, and the operational security programs. The scope of the assessment must include: (1) reference to the security assessment for either the design certification or manufacturing license, (2) a description of how security design features left unresolved by design certification or manufacturing license were addressed, and (3) scenarios that necessitate consideration of the site characteristics and the security operational programs. Ultimately, security design issues identified by this or a previous assessment, which are not resolved by a security design feature, should be identified by a security assessment parameter and must be resolved by the operational security program.

If the COL application proposes to use a custom design, then the scope of the security assessment would include a complete security assessment, including what would otherwise have been performed at the design certification stage, as described above. A COL applicant referencing an already-certified design would not be required to make enhancements to the plant design within the scope of the design certification.

If the COL application proposes to use a standard design approval, then the scope of the security assessment would include a complete security assessment, including what would otherwise have been performed at the design certification stage, as described above. A COL applicant referencing an already-certified design would not be required to make enhancements to the plant design within the scope of the design certification.

In addition to the design characterization described above, a part of the facility characterization involves a description of the operational scope that needs to be considered in the target set analysis. Therefore, each applicant should characterize the operational scope that is included in the assessment such as the plant operational modes, security alert levels and maintenance configurations, including phases of construction for multi-unit sites. Note that NRC regulations encompass all modes of operation and maintenance configurations, in accordance with the requirements in 10 CFR 73.55(f)(4) and 10 CFR 73.58, "Safety/Security Interface Requirements for Nuclear Power Reactors."

2.3.2 Perform Target Set Analysis

The security assessment process includes a target set analysis on the facility design as characterized in Section 2.3.1. The target set analysis is a systematic approach to identify complete sets of adversary targets.

With the facility characterized, the targets can now be identified. In this section, the adversary objective being analyzed is radiological sabotage that has the potential to cause radiological release. Theft and diversion of radioactive materials should also be addressed if un-irradiated MOX fuel is present on site.

A radiological sabotage target set is the combination of equipment or operator actions which, if all are prevented from performing their intended safety function or prevented from being accomplished, would likely result in significant core damage (e.g., non-incipient, non-localized fuel melting and core disruption) barring extraordinary actions by plant operators (10 CFR 73.2, "Definitions" (Ref. 16)). By identifying the adversary's objectives, target sets can be used to aid in the identification of the strategies necessary to prevent core damage, spent fuel sabotage and theft of radioactive materials, while allowing licensees and applicants the flexibility to better design their security programs with site-specific conditions in mind. Guidance and a methodology to identify and generate target sets are found in RG 5.81 (Ref. 9).

While the goal of the adversary is to disable a complete target set, the goal of the physical protection system is to protect targets with high assurance (Ref. 15).

Iterations of this security assessment process, outlined in Figure 2-1, may require an updated target set analysis. Physical modifications to the facility design (completed in Step 2 of the process in Figure 2-1) may change the targets within specific target sets and warrant an updated target set analysis.

2.4   Design Physical Protection System

Figure 2-4 depicts Step 3 of the high assurance evaluation process. The physical protection system at a nuclear power plant integrates people, procedures, and equipment for the protection of assets or facilities against theft, radiological sabotage, theft of special nuclear material, or other malevolent human attacks. The purpose of this step is to characterize the physical protection system to support the security evaluation to be performed in Step 4. Each element of the PPS (i.e., detection, delay, and response) influences its respective probabilistic measures (probability of detection (P_D), probability of interruption (P_I), and probability of neutralization (P_N), as described in Section 2.2.4). The probabilities that are evaluated in Step 4 of the process are explained in Section 2.5.

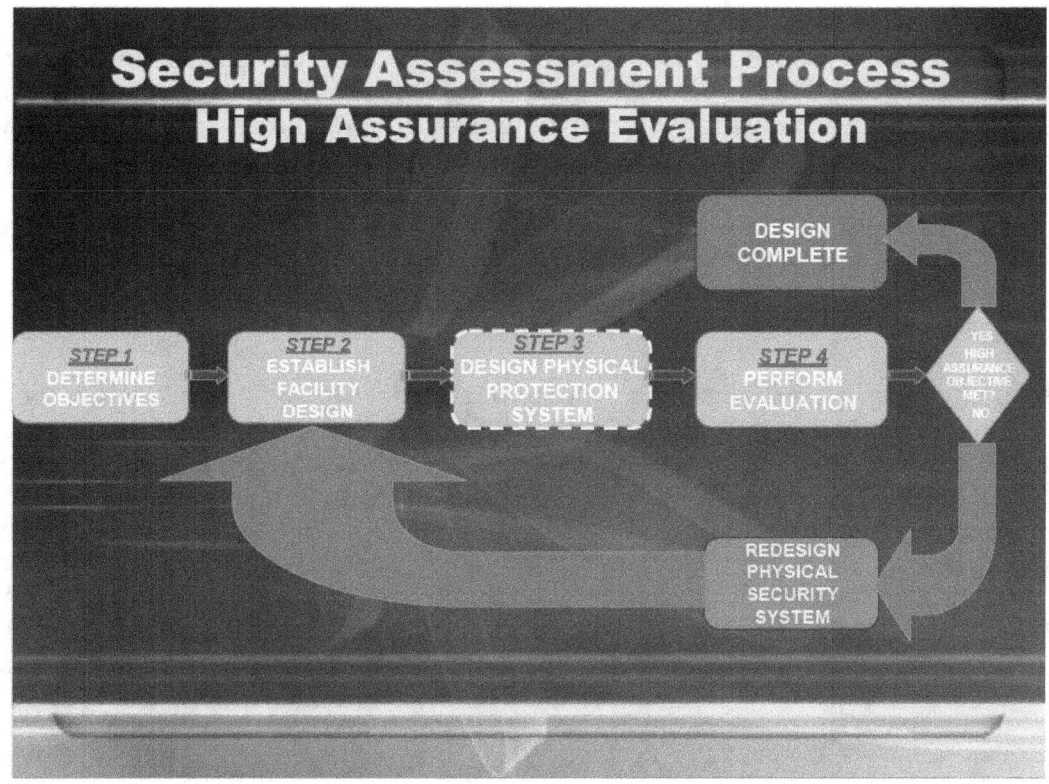

Step 3: Design Physical Protection System

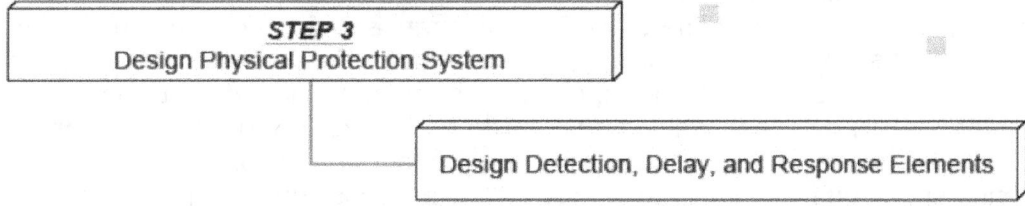

Figure 2-4 Security assessment process – step 3: design physical protection system

2.4.1   Design Detection, Delay, and Response Elements

This section describes the PPS elements that can be folded into the existing facility design, as discussed in Section 2.3, to obtain a PPS that provides high overall system effectiveness. The final design should include system functions necessary to (1) detect, delay, and respond to an attack against target sets by an adversary possessing the DBT characteristics and (2) provide conditions that facilitate mitigation actions to occur before, during, and after an attack consistent with the requirements in 10 CFR 73.55 (Ref. 1) and the guidance in RG 5.76 (Ref. 17). The applicant should identify candidate security design features that will be assessed using a risk assessment methodology to determine the effectiveness of these features in accomplishing security functions. These candidate security design features should include design concepts contained in "Nuclear Power Plant Security Assessment Technical Manual" (Ref. 7). The assessment of these features is discussed in Section 2.5.1.6.

In addition to the candidate security design features, applicants may conduct iterative runs of the security assessment process to assess different detection, delay, and response elements that could potentially be added to the design to reach the desired response margin or overall system effectiveness. Each iteration through the process (as PPS elements are added, modified, and deleted from the facility design) may produce insights about which PPS elements are the most efficient and effective (i.e., which elements cost the least while greatly enhancing physical security system effectiveness and vice versa). These insights should be captured and documented in the security assessment submission.

2.4.1.1   Detection Elements

The detection function in a PPS includes exterior and interior sensors, monitoring of barriers by security personnel (as applicable under 10 CFR 73.55(e)(8)(ii)), alarm assessment, access control, and the alarm communication and display subsystems, all working together. An effective PPS should first detect an intrusion, generate an alarm, and then transmit that alarm to a location for assessment and appropriate response (Ref. 14). The intrusion detection should occur as early as possible in the adversary task timeline. The chosen detection functions of the PPS will affect the probability of detection, which is measured both by the probability of sensing adversary action and by the time required for assessing and reporting the alarm (Ref. 18). The NRC-recommended probability of detection for a protected area perimeter intrusion detection system is 90 percent detection with 95 percent confidence, as stated in NUREG 1959, "Intrusion Detection Systems and Subsystems" (Ref. 19). Detection elements should be identified and described in adequate detail to support the security assessment. Lanuage in 10 CFR 73.55 requires the detection of both attempted and actual penetration of the protected area perimeter barrier before completed penetration of the protected area perimeter barrier to ensure that an adequate response by the security organization can be initiated. In addition, 10 CFR 73.55 requires that all vital area access portals and vital area emergency exits have intrusion detection equipment and locking devices. Therefore, high assurance is aided by intrusion paths with additional intrusion sensors. For intrusion paths through barriers without intrusion detection equipment (walls, underground pathways, etc.), information on the patrols or observation capabilities of security personnel should be detailed enough for reviewers to determine that detection would occur before exploitation with high probability and confidence. Detection and assessment by personnel should be accompanied by a discussion on communications (primary, secondary, and duress alarms if appropriate), and surveillance and assessment capabilities during different environmental conditions.

2.4.1.2   Delay Elements

An effective PPS should provide sufficient delay after initial detection of the adversary to allow time for a suitable response. The chosen delay functions of the PPS will affect the probability of interruption, which is the probability that the response force arrives at the interruption point (a pre-determined defensive position) before the adversary completes his attack sequence. Detection is desired to occur such that there is adequate time for the response force to reach the interruption point before the adversaries are beyond the effectiveness of this position. The critical detection point (CDP) is defined as the point on the adversary's path where path delay exceeds response force interruption time.[4] Delay elements should be identified and described in adequate detail to support the security assessment, to include, but not limited to: access

---

[4]   The CDP concept and use in assessing overall PPS effectiveness is described in detail in Section 2.5.1.5 of this guide.

points (gates, doors, turnstiles); barriers (fences, razor wire, walls, windows, grates, vehicle barriers); and other structures (bullet resistant enclosures (BREs), pipes, "ankle breakers," plant machinery and structures that could provide a pathway). Those delay elements of the PPS that are activated and placed after the detection of a security event should be described in detail, including activation procedure and timeline to full effectiveness of the barrier. Delay characteristics to include traversal times may be found in SAND2001-2168, "Access Delay," Volume I, Technology Transfer Manual, August 2001 (Ref. 20), Regulatory Issue Summary (RIS) 2003-06, "High Security Protected and Vital Area Barrier/Equipment Penetration Manual," March 20, 2003 (Ref. 21), and NEI 09-05, "Guidance on the Protection of Unattended Openings that Intersect a Security Boundary" (Ref. 22).

2.4.1.3   Response Elements

The response subsystem of the PPS involves two interrelated factors: the time it takes for the desired responders to arrive at the proper location and the effectiveness of that response once responders are at that location. In combination with the response capability, the response strategy should also be considered. Basic response strategies used include denial, containment, interruption, and recapture and recovery. Deterrence may be a factor, but it is usually because of a robust security posture and difficult to quantify. The chosen response strategies of the PPS will affect both the probability of interruption and the probability of neutralization. Response elements should be identified and described in adequate detail to support the security assessment. This detail should include: armament, personnel armor, BRE design (bullet and blast resistance, field of fire, communications, restricted lines-of-sight/ blind spots), response pathways, positions, and timelines for all response personnel. If remotely operated weapon systems (ROWS) are planned, all aspects of the ROWSs should be detailed in a separate report and necessary details for the purposes of the security assessment should be included (command and control, operator locations, system description, fields of view, fields of fire, system limitations, etc.).

2.4.1.4   Safety/Security Interface

The addition of PPS elements may have an effect on reactor safety. Conversely, changes in plant operation or equipment configuration may affect security. Therefore, as the PPS is designed, the applicant should consider the role of safety. This consideration is termed the safety-security interface. Safety and security interface refers to the actual or potential interactions that may adversely affect security activities because of design or operational (including maintenance) activities or vice versa. Requirements for managing the safety security interface are found in 10 CFR 73.58 (Ref. 23).

To achieve the objective of optimizing the design features and operational controls and to balance the needs of safety and security, an evaluation of any proposed PPS change using the applicable safety requirements, the probabilistic risk assessment (PRA), and a security assessment is recommended. Further guidance can be found in RG 5.74, "Managing the Safety/Security Interface" (Ref. 24).

Operational (Safety) Initiated Changes

For each proposed operational-driven change, in addition to the appropriate safety reviews, a review of the change's impact on security should be assessed. This can be done by reviewing the elements of the physical protection systems to determine which, if any, are affected by the proposed change. The type of analysis depends on the element or elements impacted by the change and the degree to which they are impacted. The evaluation can range from a simple screening analysis to an integrated benefit analysis. Therefore, before the implementation of a proposed safety change, the change should be evaluated using the security assessment process.

Examples of adverse operations initiated interactions that may have occurred at nuclear power plants include: (1) inadvertent security barrier breaches while performing maintenance activities (e.g., cutting of pipes that provided uncontrolled access to vital areas, removing ventilation fans or other equipment from vital area boundary walls without taking compensatory measures to prevent uncontrolled access into vital areas), (2) blockage of bullet resisting enclosure's (or other defensive firing position's) fields of fire, (3) erection of scaffolding and other equipment without due consideration of its impact on the site's physical protection strategy, (4) staging of temporary equipment within security isolation zones, and (5) extended maintenance outages of equipment, which are identified targets within the target set analysis.

Security Initiated Changes

For each proposed security-driven change, its impact on plant safety should be assessed. Changes made to structures, systems, and components (SSCs) that are out-of-scope of the PRA and are out-of-scope of other deterministic safety requirements that are included in the design or licensing bases of the plant can be screened. Changes made to components within the scope of the PRA or other deterministic safety requirements should be evaluated for adverse safety impact. Therefore, before the implementation of a proposed PPS change, the change should meet appropriate safety criteria.

Examples of security activities that have the potential to adversely affect safe plant operations include: security force staffing changes on backshifts, weekends, and holidays that could adversely impact operations during plant events or emergencies (e.g., opening and securing vital area access doors to allow operations personnel timely access to safety-related equipment) and the installation of security equipment that interferes with plant operations (e.g., placement of a security fence that blocks the pressure relief blowout panel for the turbine driven auxiliary feed water system and installation of security delay fencing with razor wire preventing access to plant fire hydrants).

2.4.1.5   Insider Threat

The threat of a passive or active insider should be considered in the evaluation of the physical protection system, consistent with the DBT. The use of a defense-in-depth approach can minimize the effect of an insider by providing redundancies to those physical protection system elements that are vulnerable to a passive or active insider. By omitting a single vulnerable element (e.g., as an armed responder, an alarm station, an intrusion detection sensor) at a time, the defense-in-depth nature of the physical protection system can be determined and improved, and thus better protected against an insider threat.

2.4.1.6   Other Site Security Design Features

Included for consideration should be those design features that enhance the effectiveness of those PPS elements that detect, delay, and respond to threats. These features can be a part of the original facility design, or added after the evaluation as part of the PPS redesign in Step 6. Examples include serpentine or channeling barriers, and closed-circuit television systems not used for assessment at the isolation zone.

2.5   Perform Evaluation

Figure 2-5 depicts Step 4 of the high assurance evaluation process. The evaluation, performed as part of the security assessment process, is intended to demonstrate that a reactor facility, including the PPS design, provides high assurance of compliance with 10 CFR 73.55 and high overall system effectiveness. The evaluation will enable NRC staff to determine whether the applicant has provided defense in depth through the integration of systems, technologies, programs, equipment, supporting processes, and implementing procedures to ensure that the capabilities to detect, assess, interdict, and neutralize threats up to and including the design-basis threat of radiological sabotage are maintained at all times.

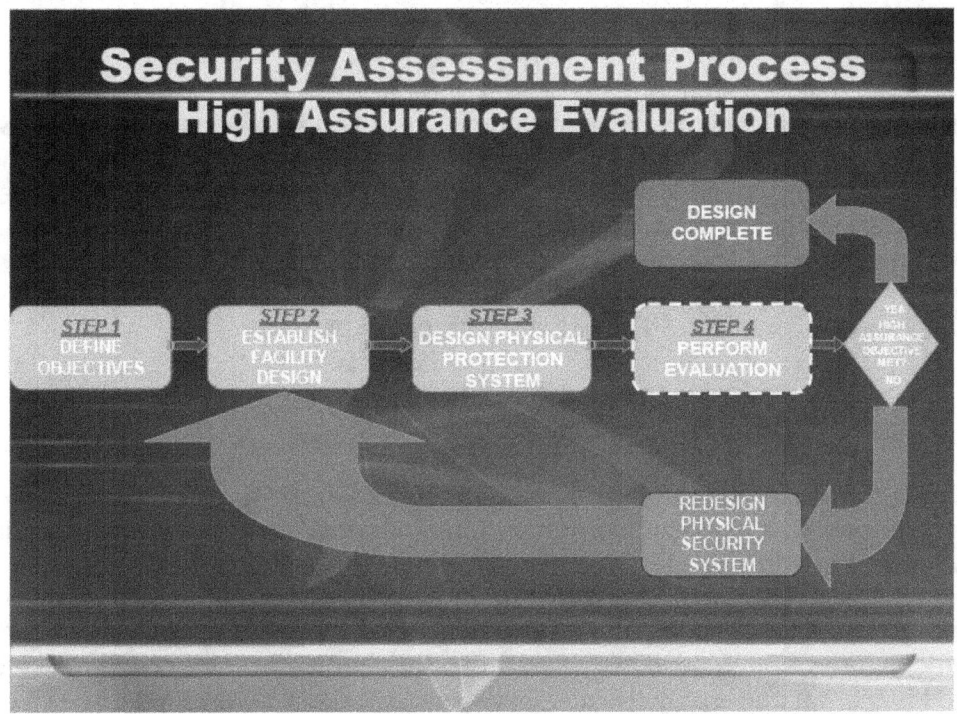

Step 4: Perform Evaluation

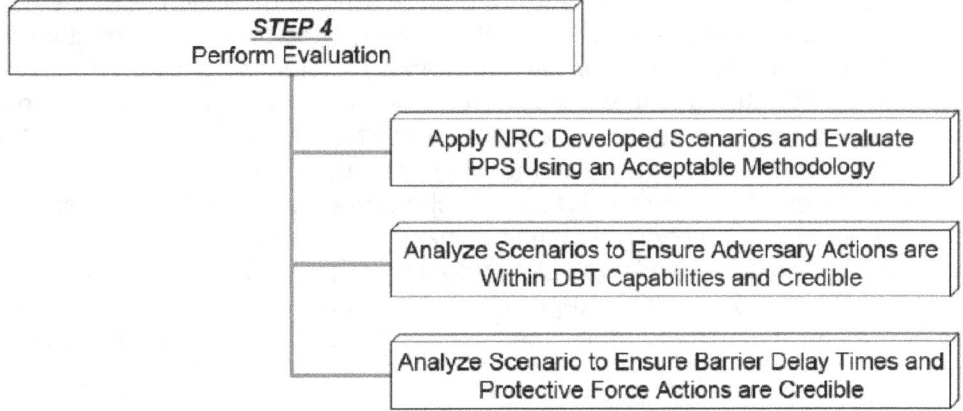

Figure 2-5 Security assessment process – step 4: perform evaluation

Methods of evaluation that are acceptable include, but may not be limited to documented table-top analysis (using Vulnerability Integrated Security Assessment (VISA) Manual and pathway analysis (using Analytic System and Software for Evaluating Safeguards and Security (ASSESS)) with Joint Conflict and Tactical Simulation (JCATS) (Ref. 25). The Estimated Adversary Sequence Interruption (EASI) model addresses all of the analysis elements with the exception of the armed protection force's ability to stop an attack. These models are described in Appendix C. Regardless of the method used, a pathway analysis, to determine probability of interruption, and a conflict analysis, to determine probability of neutralization, should be performed.

2.5.1 Apply NRC Developed Scenarios and Evaluate Overall PPS Effectiveness Using an Acceptable Methodology

The NRC staff will provide a standard set of scenarios associated with the DBT (Ref. 3) that defines basic characteristics of the adversary force, including force size, equipment, weapons, and tactics. These standard scenarios are the basis for developing adversary timelines and blast effects analyses. Each standard DBT scenario may result in several overall scenarios that vary based on entry points, target sets, timeline analysis techniques, protective force response, etc. The combination of the DBT scenario, target set, and entry and exit points (exit points for theft scenarios) help to define the adversarial pathway that contributes to an attack. Other factors, such as the design of the PPS, will also affect the pathway.

Application of the scenarios using the methods described below can produce either a qualitative or quantitative assessment of overall PPS effectiveness.

The qualitative assessment seeks to determine if there is adequate margin in the time it takes for the protective force to either reach predetermined protective positions with weapons at-the-ready, or for the protective force to activate denial and delay systems, or both, before the adversary disables all the targets for any given target set. As the times used in the assessment are average or mean values, a margin on the order of one standard deviation in the distribution of the differences between the adversary and protective force timelines would indicate high assurance. This margin addresses the most likely variations in adversary and protective force timelines and is necessary to meet the intent of the overall security objectives of preventing significant core damage, spent fuel sabotage, and theft and diversion of radioactive materials.

The quantitative assessment uses a measure of overall system effectiveness to evaluate the objectives of preventing core damage, spent fuel sabotage, and theft of radioactive materials. This overall system effectiveness is determined in part by evaluating the P_I and the P_N, as shown in Equation 1 of Section 2.2.4, where interruption is defined as arrival of responders, or activation of denial systems at a deployed location, to halt adversary progress, and includes the detection, delay, and response elements of the PPS. Neutralization is defined as the defeat of the adversaries by the protective force. Detection elements and the adversary and protective force timelines are used to quantify interruption. The P_N is calculated as the measure of the likelihood that the protective force will be successful in overpowering or defeating the adversary, given interruption. A discussion of the range of values that can be associated with system effectiveness and their relationship to the objective of high assurance is included in the following subsections.

The basic steps for the determination of the overall PPS effectiveness include:

- identify the overall scenarios

- evaluate blast effects
- analyze scenario timelines
- analyze neutralization
- determine overall PPS effectiveness (integration of the first four elements)
- risk-informed evaluation of candidate design features

Each of these steps is described in the following sections.

2.5.1.1 Overall Scenario Identification

At a given facility, each unique combination of DBT scenario, entry points, exit points, adversary pathways, protective force response, and target sets defines an overall attack scenario. For radiological sabotage, only the entry paths are evaluated; for theft or diversion, both entry and exit paths should be evaluated. The objective of scenario identification is to identify the set of overall scenarios that will be used to determine the effectiveness of the PPS. This section uses a simplified example to illustrate this process.

Assume a security assessment is being performed on the facility shown in Figure 2-6. This simplified facility contains two target sets (Target Set A and B) and has a protective force of two, with two designated tactical positions or CIPs (Critical Interruption Points).

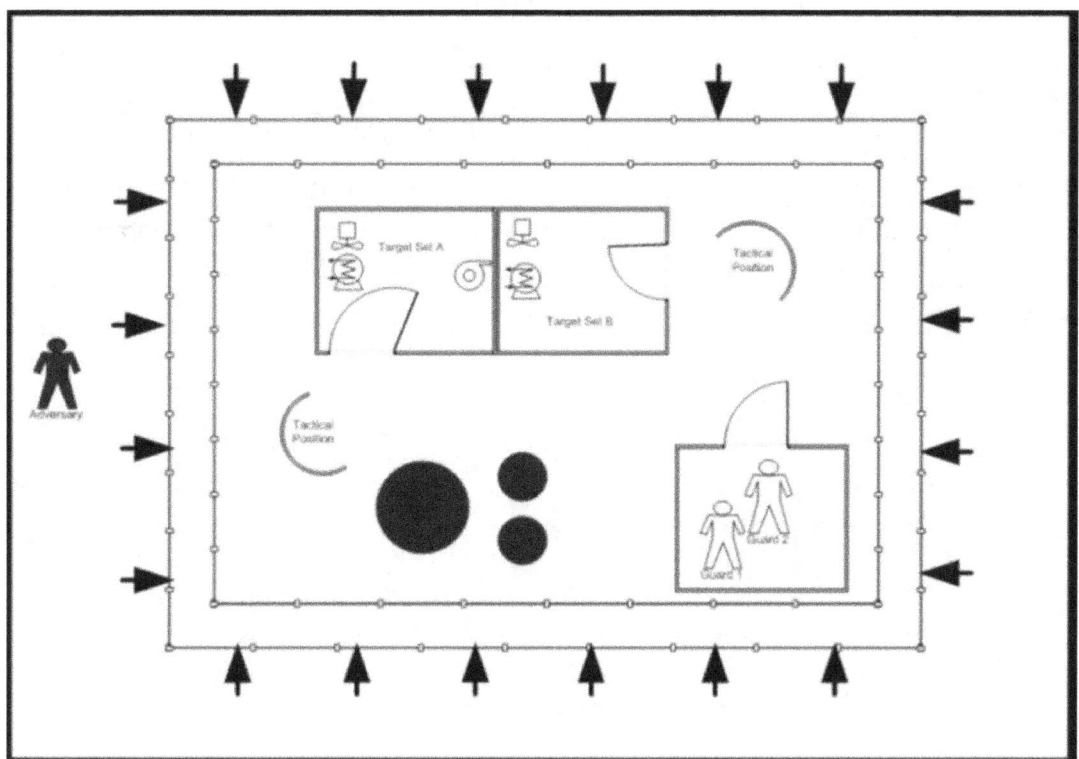

Figure 2-6 Example facility

2.5.1.2 Blast Effects

Blast effects should be determined for the specific scenario as described in the standard set of NRC scenarios and the defined overall scenario. A discussion of blast effects is now found in

RG 5.68, "Protection Against Malevolent Use of Vehicles at Nuclear Power Plants" (Ref. 26) and RG 5.69 (Ref. 12).

In addition, assume the DBT scenario, DBT Scenario #1, is as shown in Table 2-1 (used for illustration purposes only—not related to characteristics of an actual DBT scenario).

Table 2-1 Example DBT Scenario #1

Attribute	Characteristic
Attack type	Sabotage
Number of adversaries:	1
Weapon load weight carried by each adversary	[Based on adversary tactics and objectives]
Body armor	Yes (UL Level 4)
Weapons	Automatic, AK-47; Ammo: 7.62mm
Terrain type	Land-based
Tactic	Overt
Transportation method	Pedestrian
Entry points	1
Exit points	N/A (not theft or diversion)
Explosive load per adversary	[Based on adversary tactics and objectives]
Insider information obtained	Target set equipment and operator action(s) locations
Cyber challenges	None

As seen in Figure 2-7, the adversary can enter the facility from any point along the fence perimeter and still be consistent with the DBT scenario of one adversary entering at one location. Therefore, the DBT scenario only establishes the characteristics of the threat, not of the overall attack scenarios. An assessment of the DBT scenario and the facility, including the PPS, should be performed to determine the overall attack scenarios.

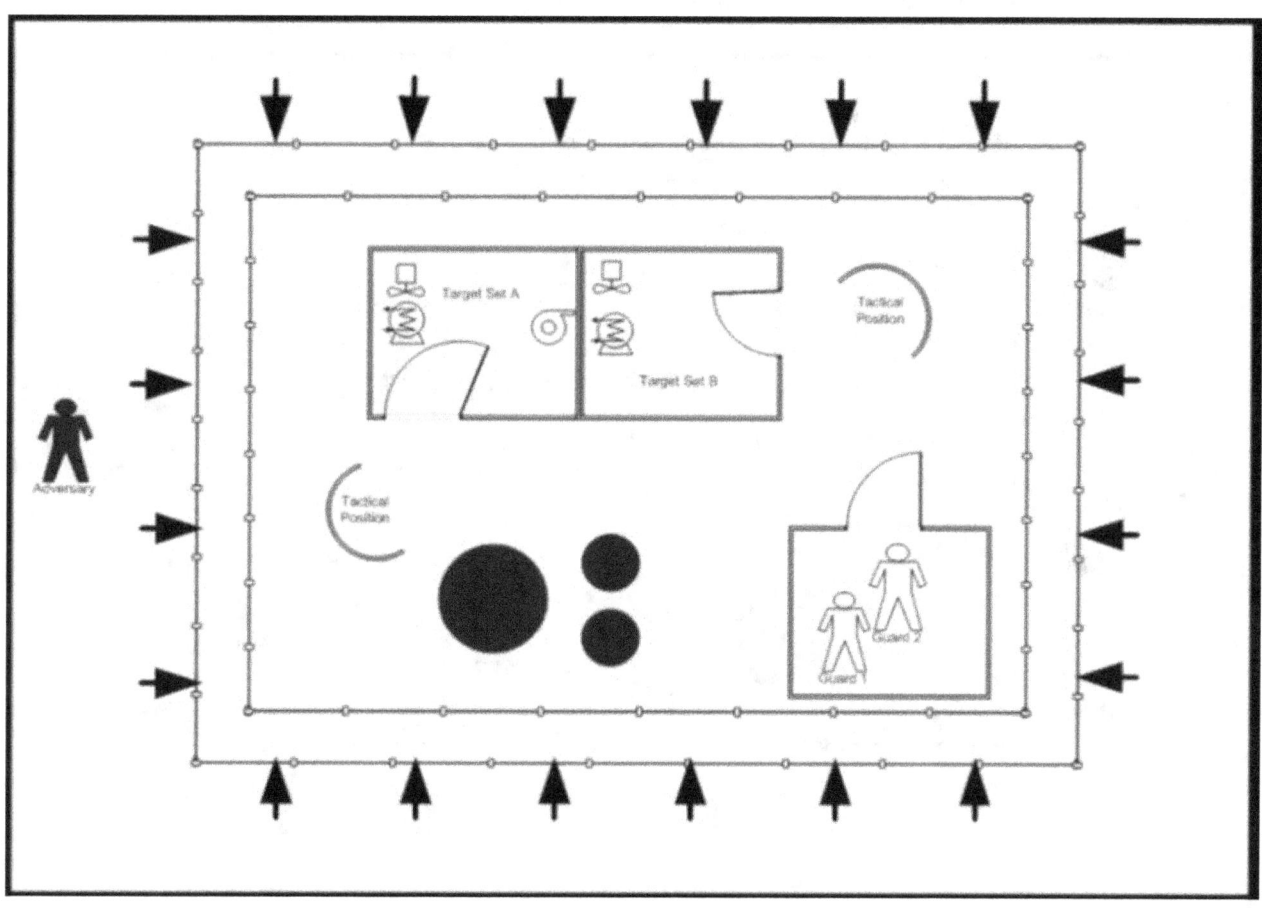

Figure 2-7 Potential adversary entry points

Each entry point creates a potential starting point for one or more overall scenarios. From these entry points, adversary pathways can be postulated for each achievable target set. The objective is to identify potential pathways from the perspective of the adversary. Consideration should be given to target access points, detection devices, travel distances, protective features, and anticipated protection force routes. Figure 2-8 shows examples of potential scenarios for Target Set A and Figure 2-9 shows scenarios for Target Set B.

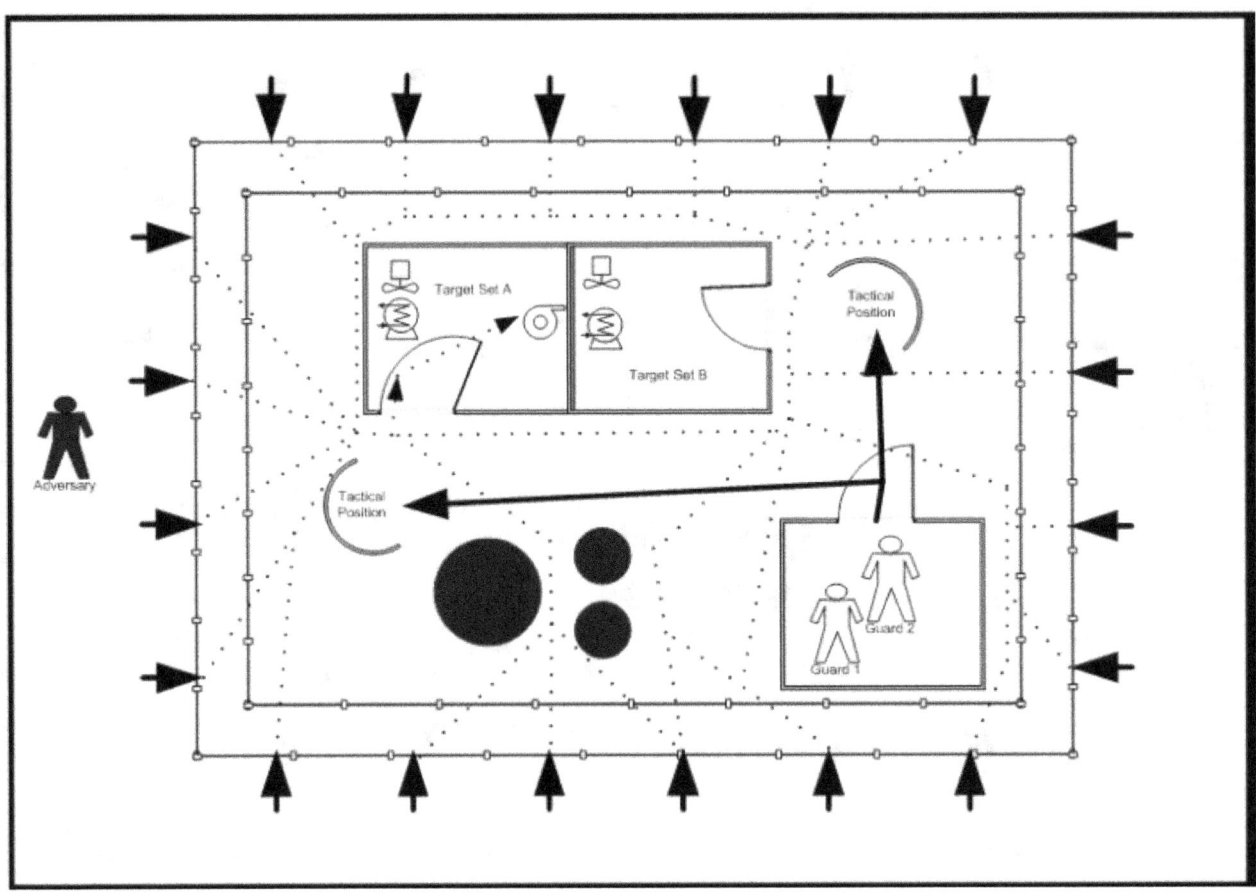

Figure 2-8 Potential adversary paths for target set A

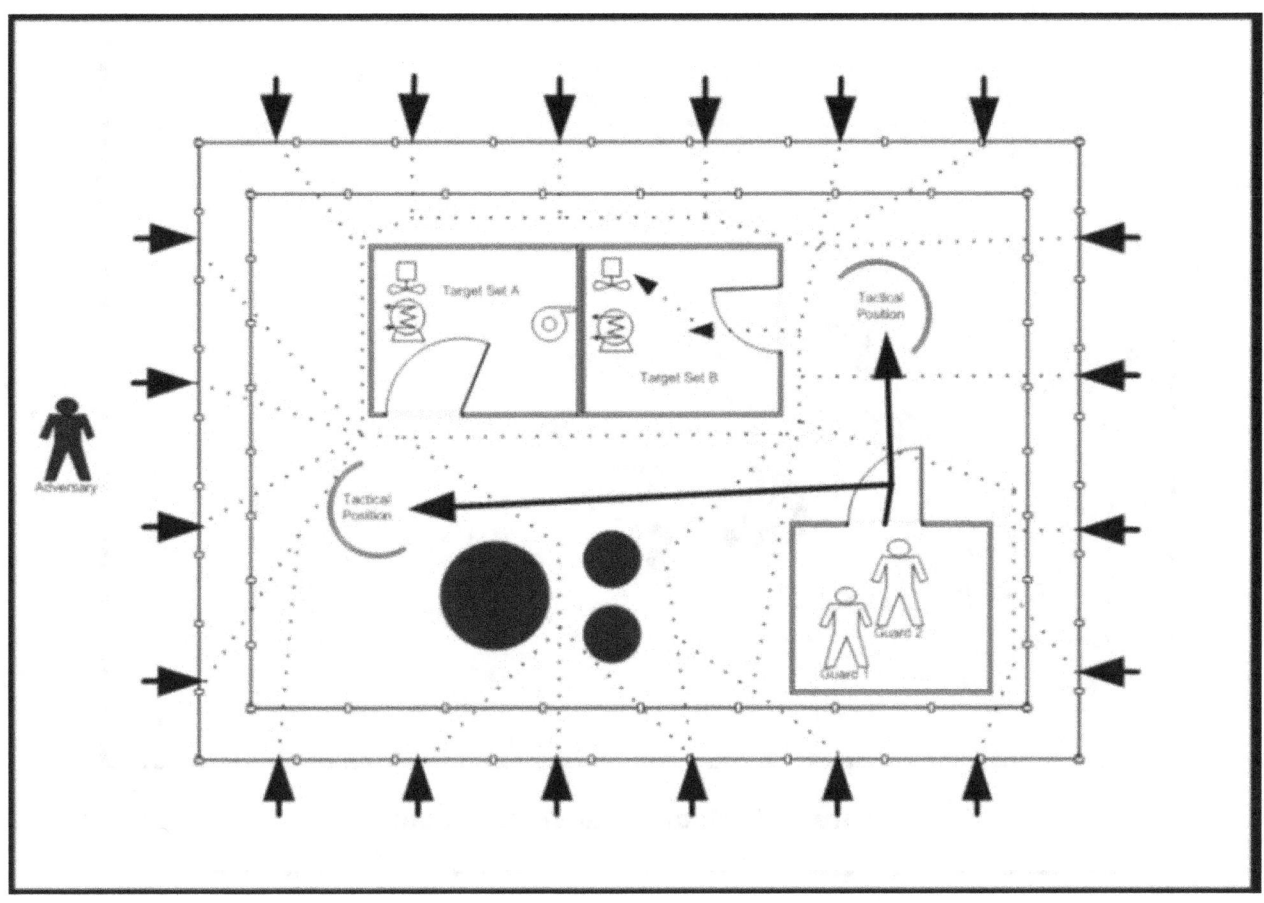

Figure 2-9 Potential adversary paths for target set B

It should be noted that in the determination of the overall scenario, both adversary and protective force pathways need to be determined. In this example, only one protective force starting point and response is anticipated. It was also determined (assumed for this example) that this is a bounding response and is consistent with the security plan.[5] If multiple security responses are possible, each would need to be assessed. These multiple responses could be evaluated through detailed scenario analyses that evaluate each response to each target set and adversary pathway. They could also be assessed through a process that identifies one or more bounding responses that maximizes distance, exposure, and response timing of the protective force by establishing the worst case response that remains consistent with the constraints of the security plan.

Even with minimizing the protective force responses, there is a potential for a large number of adversary paths based on different entry points, target sets, and defensive features. To manage the security assessment, a process can be used to determine the most vulnerable path(s) (those with the lowest P_I). These may be the shortest paths (paths that minimize adversary time to the target) as highlighted in Figure 2-10 or paths that provide the adversary with tactical advantages as shown in Figure 2-11.

---

[5]   For design certification applicants, consistency with the security plan does not apply.

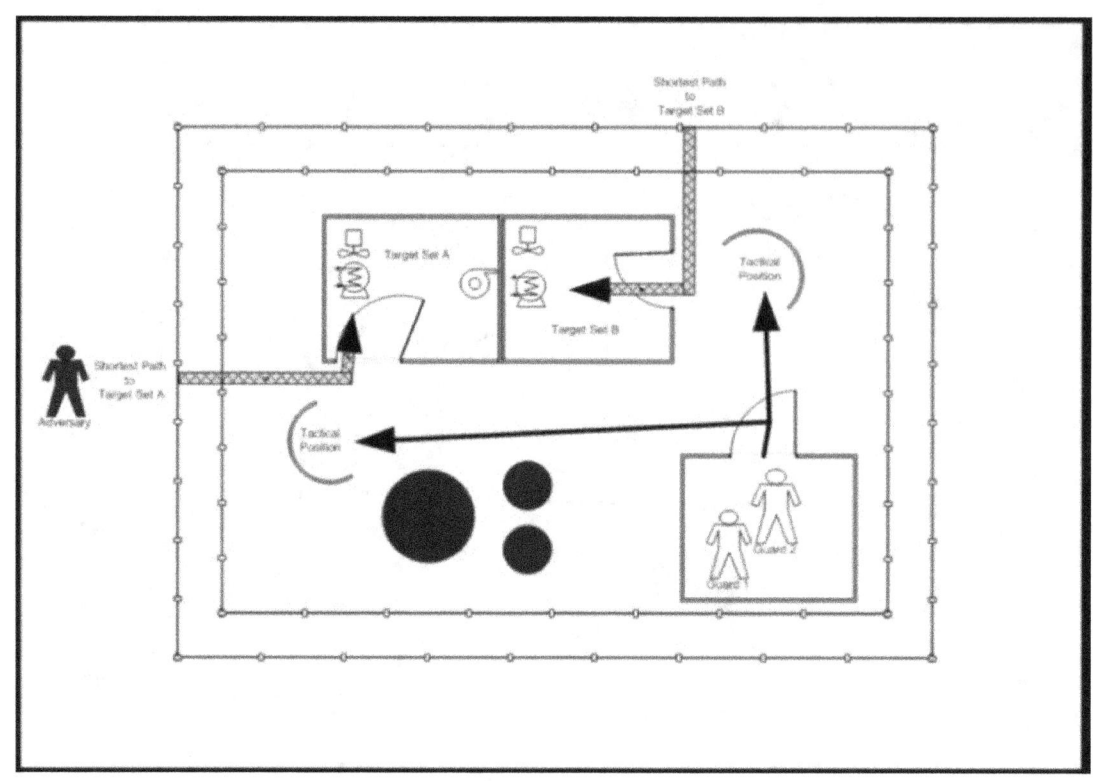

Figure 2-10 Shortest adversary paths

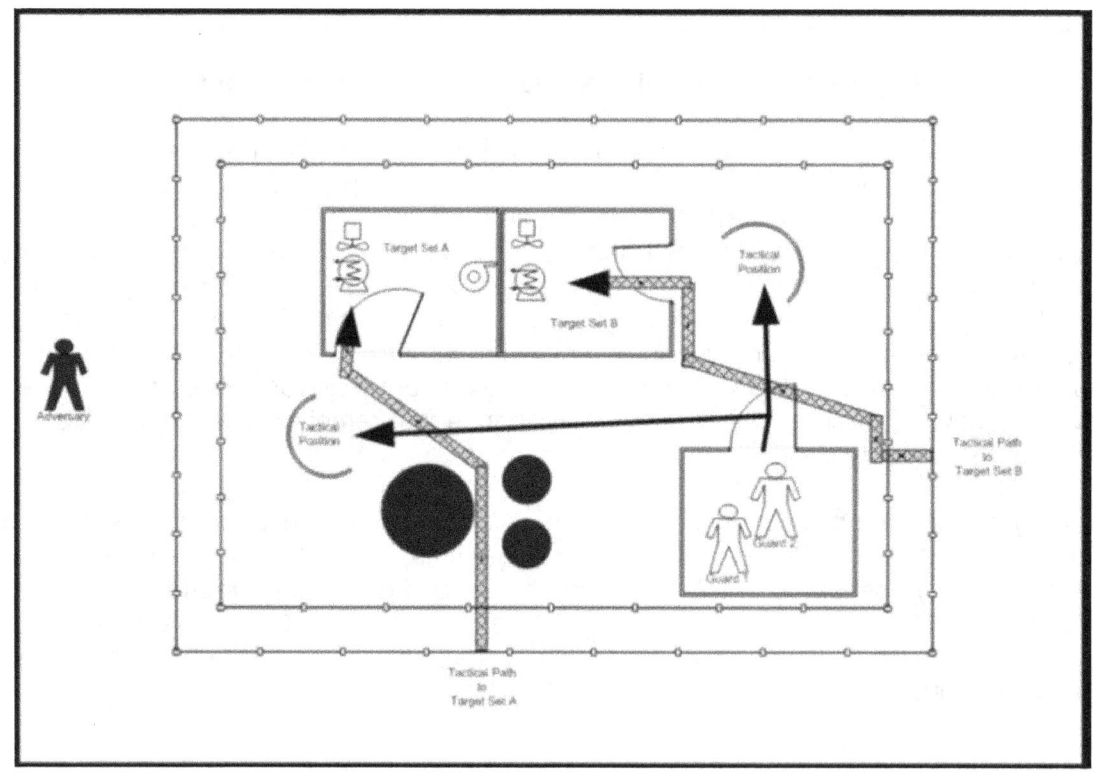

Figure 2-11 Potential adversary paths with tactical advantage

As illustrated by Figure 2-11, the most advantageous paths to the adversary may not be the shortest paths. Paths where the adversary may have an enhanced tactical position could be paths that give the adversary the ability to engage the responders before reaching their CIPs or that take advantage of cover provided by plant structures that may reduce the effectiveness of the pre-established tactical positions. Paths that minimize detection and combination paths that minimize detection until a likely detection point is reached and then minimize time also could be identified. The most advantageous paths may be through barriers, walls or structures, as the adversary may use equipment or explosives to create new entry points. A breaching analysis of protected area barriers, vital area barriers, and other barriers or delay features, serves an important role in the PPS. The objective of the overall scenario identification process is, for a given DBT scenario, to identify the overall scenarios that provide the most advantages to the adversary. If high assurance can be demonstrated for these most challenging bounding scenarios, then high assurance may also be demonstrated for lesser challenges.

At this stage in the analysis, the identification of these limiting (or critical) overall scenarios could be done through the use of a documented table-top analysis (using VISA manual), a pathway modeling analysis tool (such as ASSESS), or by documenting hand calculations. ASSESS is described in Appendix C. If a table-top analysis is performed, the methods and assumptions used to identify the limiting overall scenarios need to be clearly stated. ASSESS can determine the critical path for the adversary to take through the facility, which minimizes the adversary's time to target or minimizes the probability that the adversary will be detected. It can search quickly through thousands or millions of paths, using different operating conditions, threats, and targets. The Adversary Time Line Analysis System (ATLAS) is an improved version of ASSESS and could be used as well (see Appendix C).[6]

In the example, it is assumed that four overall scenarios are identified through the scenario assessment process. These scenarios are shown in Table 2-2.

Table 2-2 Resulting Overall Scenarios

Overall Scenario	1-1	1-2	1-3	1-4
DBT scenario	#1	#1	#1	#1
Target set	A	A	B	B
Type	Short (minimizes time)	Tactical (minimizes detection)	Short (minimizes time)	Tactical (minimizes detection)
Entry point	West Fence - Midpoint	South Fence - Tank Farm	North Fence - Target Set B Entrance	East Fence - Office Complex

The overall scenarios identified in Table 2-2 are examples of the results that could be obtained from a process that limits the number of scenarios analyzed by identifying those that provide the

---

[6] The assessment and modeling-type tools used throughout this NUREG and described in detail in Appendix C are for illustration purposes only. The NRC does not endorse any specific assessment or modeling-type tool for security assessments.

greatest advantage to the adversary. These scenarios would be candidates for the timeline analysis discussed in the next section.

2.5.1.3   Scenario Timeline Analysis

Scenario timelines should be developed and analyzed for each of the most challenging bounding scenarios.

<u>Adversary Timeline</u>

The adversary timeline is an assessment of the impact of the physical protection functions of detection, delay, and response on the adversary for a given DBT scenario, given specific entry points (and exit points, if applicable) and a given target set. Each unique combination of DBT scenario, entry points, and target set defines an adversary sequence diagram. For radiological sabotage, only the entry paths are evaluated; for theft or diversion, both entry and exit paths should be evaluated. Figure 2-12 shows an example of an adversary timeline for radiological sabotage.

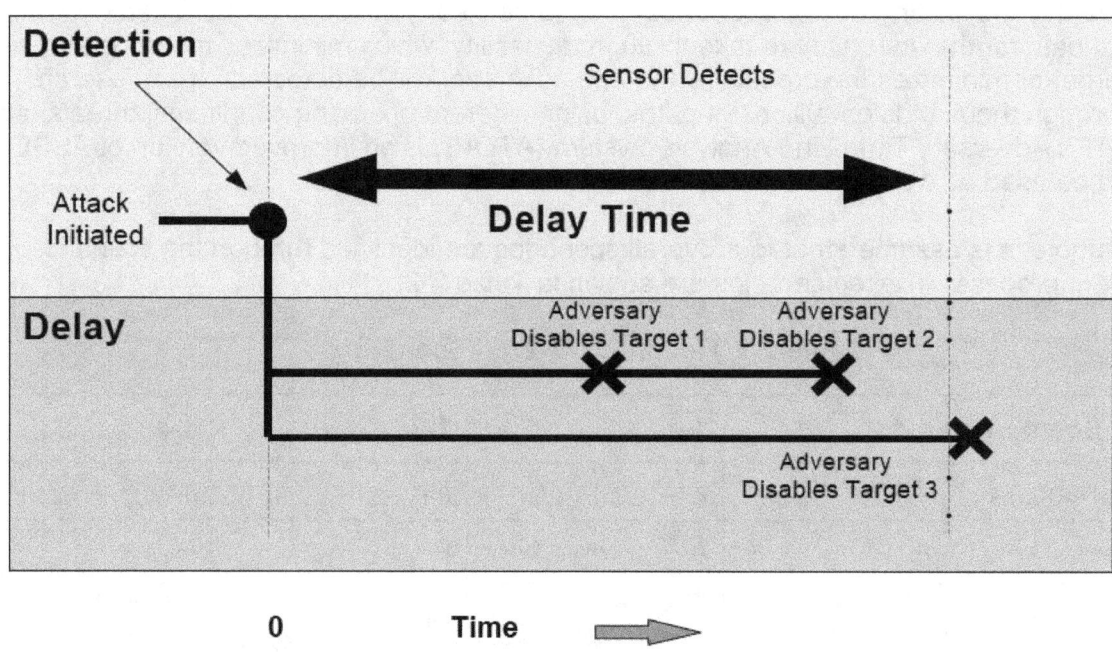

Figure 2-12 Adversary timeline

Figure 2-12 shows the adversary timelines for disabling a three element target set with two groups of adversaries. Note that delay time is not initiated until a sensor detects the presence of an adversary and terminates when all targets are disabled. Delays before detection should be excluded from the adversary timeline because of the lack of an initiating cue to call the protective force to action. Delays after detection can be associated with passive delay barriers or activated delay systems. As it is expected that for any given adversary path that there will be multiple detection opportunities (regulations require protective area detection and vital area detection), the detection point used for establishing a scenario timeline is the one that has a high probability of detection or high accumulated probability of detection and that yields the most effective (i.e., the one with the most margin) response timeline from among the credited detection opportunities.

The adversary timelines can be depicted with several different approaches: logic diagrams, event trees, and adversary or sequence diagrams. All event sequences should be diagramed from the perspective of the adversary as a tactical map of activities and events necessary to achieve its objective. The entire target set should be addressed for a radiological sabotage objective to be complete. This may require activities to be accomplished in parallel, requiring a more complex diagramming tool. When creating the timelines, the assumptions associated with the various design features (i.e., assumptions associated with the effectiveness of detection and delay features) should be clearly stated.

Segmentation

Segmentation of the complete pathway may aid the analysis. Pathways are typically composed of multiple segments or a subset of events that contribute to an attack. In the earliest stages of development, the assessment can be organized in coarse pathway diagrams that serve as the basis for judgmental quantification. As more design detail becomes available, more detail is added to the pathways through pathway segments, and engineering analysis replaces judgment in assessing the probabilities and measures (Ref. 27).

Protective Force Timeline

The protective force timeline is an assessment of the time it will take for one or more members of the security force to reach a location or activate a system where an adversary's path can be interrupted. This timeline should consider (1) the time it takes the detection signal to process from the sensor to the data gathering panel to the alarm communication and display, (2) the time necessary for the alarm communication and display annunciation to be acknowledged and assessed by the operator and (3) the time taken by the operator to communicate the alarm to the protective force or activate the delay and denial system. At the point of communication, the timeline shifts to the response phase. This phase considers the time it takes for the protective force to respond to their designated positions and/or ready weapons or for the protective force to activate delay and denial systems. Figure 2-13 shows the basic timeline for the protective force.

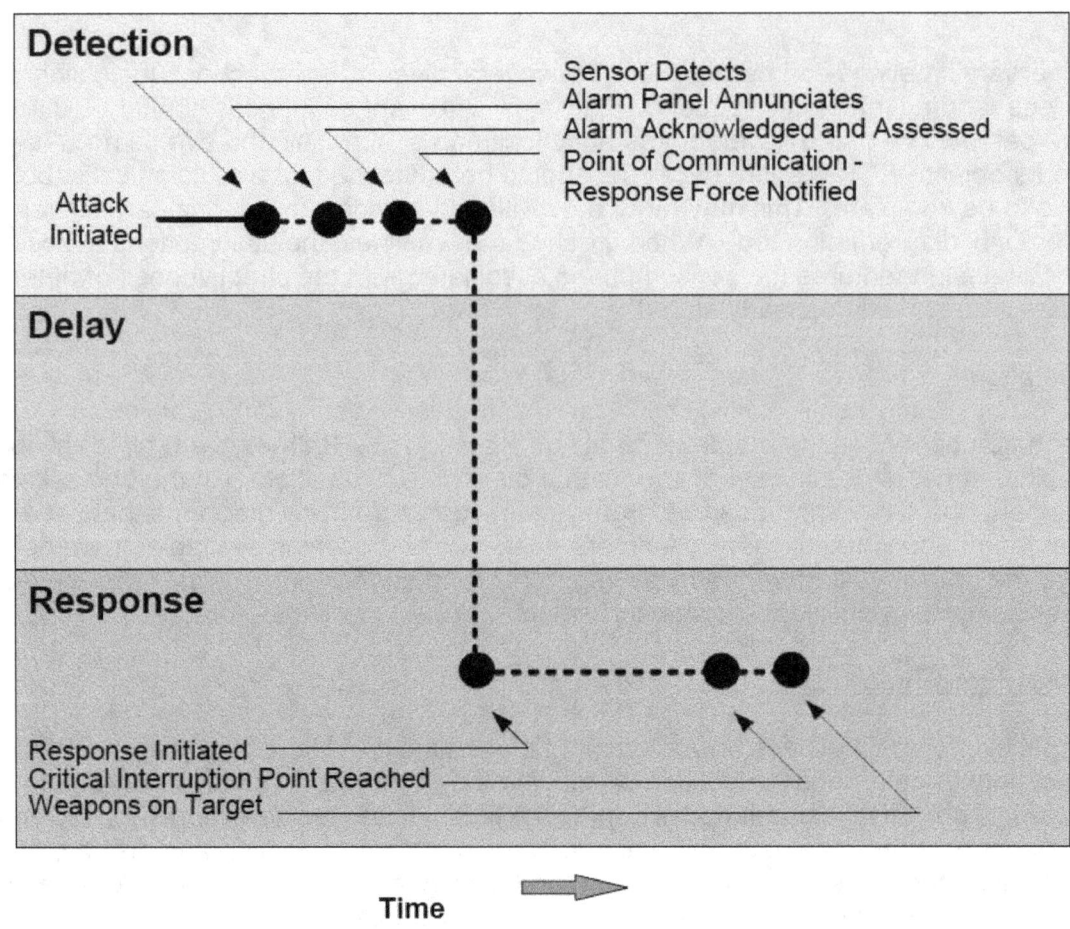

Figure 2-13 Protective force timeline

Note that the CIP is a predetermined protected location or the location of remotely operated delay and denial systems that provides tactical and strategic advantage to the responding protective force to protect one or more targets. Failure to reach the CIP or activate the systems before the adversary reaches the final target of a target set significantly reduces the likelihood of successful adversary neutralization.

The response timeline should also consider the impact of a cyber attack—if included in the DBT scenario being analyzed—on the security detection and communication systems and its effect on the detection and assessment of and the protective force response to the adversarial attack. Cyber attacks may also include those that disable a target within a target set. Treatment of cyber attacks in target sets is further described in RG 5.81, "Target Set Identification and Development for Nuclear Power Reactors" (Ref. 9). Additionally, for attack external assault scenarios that include vehicle or boat bombs, the response timeline will need to account for the required standoff (or minimum safe) distance and blast effects. Protection from blast effects is primarily accomplished by keeping the explosive source at a distance from the target. This distance is referred to as standoff distance. The amount of standoff distance required to provide an acceptable level of protection to equipment, personnel, and systems is a function of the quantity of explosives considered and the type of barriers or structures considered, if any. For further discussion, see the revised NUREG/CR-6190, "Protection against Malevolent Use of Vehicles at Nuclear Power Plants.".

2.5.1.4   Neutralization Analysis

Neutralization analysis is the evaluation of the armed protection force's ability to stop the attack once the protective force has reached an interruption location or reached a location where denial and delay systems may be activated. The objective of the protective force should be to reach or activate systems at the appropriate CIP. Timelines that show the protective force reaching or activating systems at the appropriate CIP with time available to ready weapons or activate systems and engage the adversary with an adequate time margin demonstrate high assurance.

2.5.1.5   Overall Physical Protection System Effectiveness

The assessment of the overall physical protection system effectiveness is determined for each overall scenario. The following attributes are considered in the evaluation of the overall scenario:

- the DBT scenario (radiological sabotage or theft and diversion)

- explosive device blast effects (if applicable)

- an achievable target set

- an adversary timeline (including exit pathway for theft and diversion)

- the probability of detection and communication

- a protective force timeline; probability of interruption

- the probability that an adversary is neutralized given an adversary's pathway is interrupted

The measure of the overall system effectiveness assesses the probability that an adversary will be prevented from disabling all targets within a target set to successfully perform radiological sabotage (or theft and diversion of special nuclear materials). If the actions required to complete the pathway are within the resources and capability of the adversary, then the probability of stopping the adversary depends on the capability of the PPS to:

1.   detect the unauthorized actions of the adversary,
2.   delay the adversary,
3.   interrupt the adversary, and
4.   neutralize the adversary before the task can be completed.

Defense in depth is demonstrated by effective redundancies to the four actions above for each scenario. A method of evaluating defense in depth is the omission of a PPS element from a scenario and determining the effectiveness of the PPS to protect against the DBT.

Figure 2-14 shows an integrated adversary and protective force timeline for a single target. In this figure, the protective force is shown to reach the CIP and achieve weapons on target before the adversary disables the target.

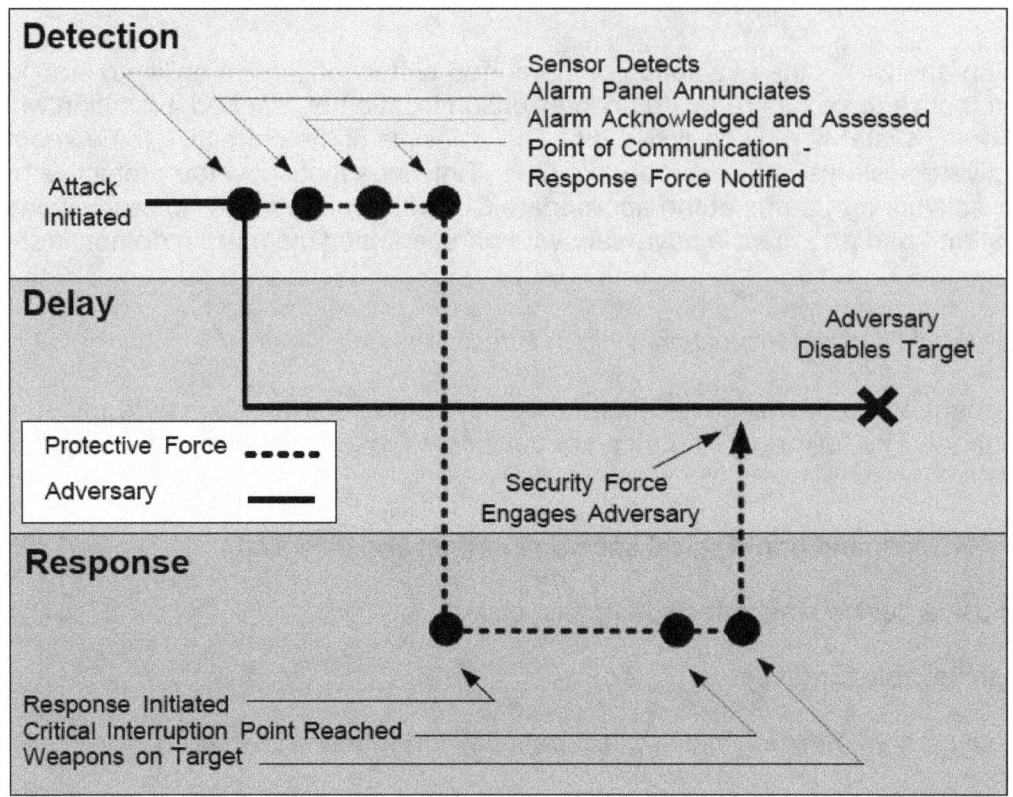

Time

Figure 2-14 Integrated protective force and adversary timelines

A typical approach to analyzing the same timeline is to shift the protective force timeline to the right until the response force engages the adversary force just before the adversary disables the target. The shifted response force timeline now begins at the CDP. The CDP is the point on the path where path delay exceeds protective force interruption time with enough time margin to allow for a high probability of neutralization. This point is found by starting at the end of the adversary path and adding up protective force path delays until this value just exceeds protective force time with an adequate margin. As seen in Figure 2-15, an adequate margin is available, providing high assurance that the protective force can prevent the adversary from achieving its objectives.

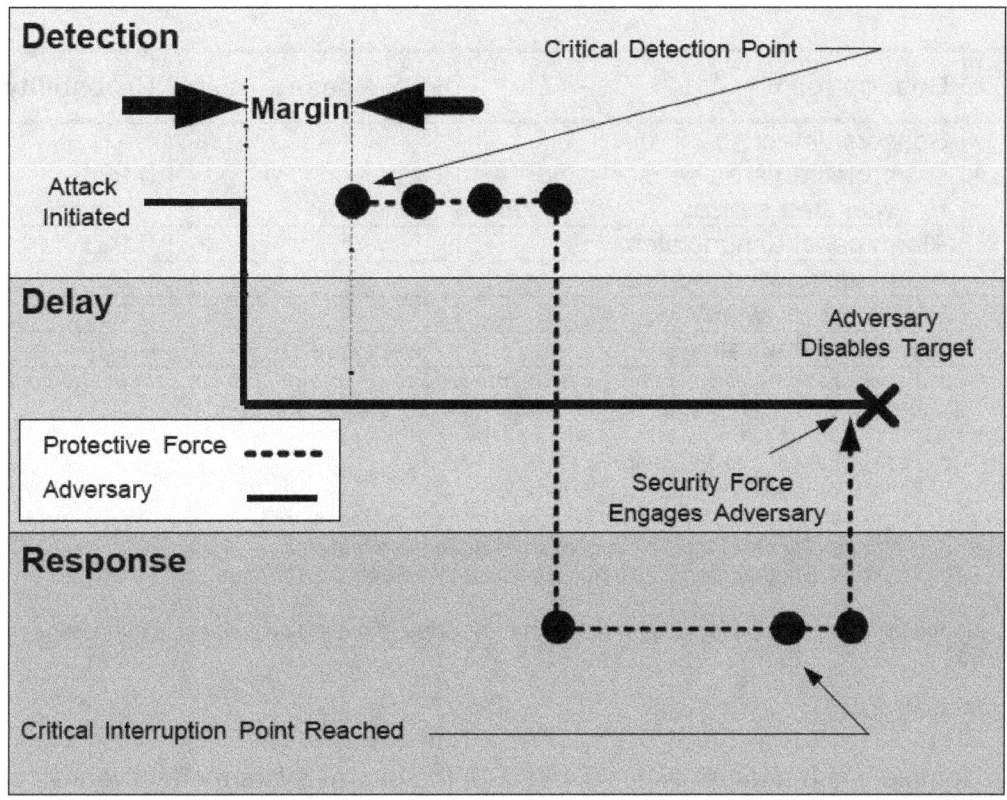

Time

Figure 2-15 Timeline margin

The CDP approach allows the response timelines to be established independent of all the adversary timelines. For each target or target set, one or more limiting response timelines can be established and these timelines can be compared against the bounding (i.e., worst case or shortest) adversary timeline that challenges the target or target set of interest. If the response timeline began at the detection point associated with each adversary timeline and an adequate margin were to be assessed at the end of the response timeline, then a response timeline would need to be constructed for each adversary timeline. Therefore, the CDP approach reduces the number of response timelines by relating the timelines to the target or target set in conjunction with the bounding adversary timeline associated with each target or target set.

This same scenario can be addressed quantitatively using Equation 1 discussed in Section 2.2.4. To use this equation, each function in the above timeline can be assigned a probability enabling the determination of the overall system effectiveness. Table 2-3 provides an example of a set of functions with their associated success probabilities.

Table 2-3 System Effectiveness Inputs

Timeline Step	Description	PPS Phase	Probability
1	Sensors detect - protected area sensor = .88 - vital area sensor = .88	Detection (P_I)	0.986*
2	Alarm panel annunciates		0.99
3	Alarm acknowledged and assessed		0.95
4	Protective force notified		0.99
5	Adversary neutralized	Response (P_N)	0.90

Sensor detection success probability is equal to the combined likelihood that both sensors fail (protected area sensor and vital area sensor). This probability is determined using the following equation:

$P_D = 1-((1-P_{D1})(1-P_{D2}) ... (1-P_{DN}))$ where
P_D = total probability of successful detection
P_{D1} = probability that Sensor 1 successfully detects an intruder
P_{D2} = probability that Sensor 2 successfully detects an intruder
P_{DN} = probability that the Nth sensor successfully detects an intruder

Therefore, the following calculation is used to determine the value of overall probability of successful detection for this example:

$P_D = 1- ((1-0.88)(1-0.88)) = 0.986$

Using the example information shown in Table 2-3, the overall system effectiveness can be determined for an overall scenario, consisting of a unique combination of a standard DBT scenario, entry points, exit points (applicable for theft or diversion), protective force response, and a target set. This example assumes that the response time is adequate (i.e., interruption is before the CIP). The probability to detect the adversary is equal to the product of the probability that the credited sensors detect an intruder, the probability that the alarm panel annunciates, the probability that the alarm station operator acknowledges and assesses the alarm and the probability that notice is given to the protective force. People, such as responders, can be credited as detection sensors, as appropriate. Note that if the adversary is able to disable both the central alarm system and secondary alarm system in one act, the probability of detection will likely be small. In this example it is assumed that there is not a single act that can disable both the central alarm station (CAS) and secondary alarm station (SAS). It also assumes that the response time is adequate (i.e., assumes interruption before the CIP). Therefore, the probability to interrupt is:

$P_I = 0.986 \times 0.99 \times 0.95 \times 0.99 = 0.92$

The protective force does not necessarily know the adversary's target set strategy for a particular attack; therefore, it needs to be sized such that it can address a range of potential targets. The goal of the adversary is to disable at least one target set while the protective force must defend all the target sets, potentially requiring several CIPs to be manned.

It should be noted that the ability to neutralize the adversary depends on the characterization of the interruption. Failure to interrupt clearly results in a zero probability to neutralize prior to the CIP. Failure to interrupt the adversary prior to the CIP reduces the protective force's tactical advantage and would decrease the 0.90 value in line 5 of Table 2-3.

In the example, interruption occurs before the CIP with weapons on target and margin. Using Equation 1, the overall system effectiveness can be determined as follows:

$P_E = 0.92 \times 0.90 = 0.83$

For a specific PPS and a specific threat scenario, the most vulnerable path (the path with the lowest P_I) can be determined. Using P_I as the measure of path vulnerability, multiple paths can be compared and, when used with P_N, an estimate of overall PPS effectiveness can be made.

Computer models such as the Estimated Adversary Sequence (EASI) model address all of these elements with the exception of the armed protection force's ability to stop an attack (P_N). Therefore, its final output can be equated to P_I, the probability of interruption. Other models exist, such as Joint Conflict and Tactical Simulation (JCATS), which, through simulated force-on-force engagements, evaluate the probability of neutralizing the adversary, if multiple runs are performed. Together, these two probabilities determine the probability of overall system effectiveness. These modeling tools are described in Appendix C.

To ensure that the measure for overall system effectiveness is valid when evaluating the facility and PPS design, the applicant should account for the validity of the adversarial action (e.g., traversal times, actions are within DBT capabilities) and PPS element assumptions (e.g., detection probabilities and delay times are correct) used in the model.

In addition to high assurance that is demonstrated through the PPS evaluation process described above, it is also necessary to demonstrate that no single act can disable the operability of the CAS and SAS in such a manner that would preclude meeting the objective of high assurance.

2.5.1.6   Evaluation of Candidate Design Features

As stated in Section 2.4.1, the applicant should identify candidate security design features that will be assessed using a risk-informed methodology to determine the effectiveness of these features in accomplishing security functions. These candidate security design features should include design concepts contained in Chapters 4 and 5 of the "Nuclear Power Plant Security Assessment Technical Manual" (Ref. 7).

This security feature evaluation should use a screening process with the goal of optimizing the inclusion of security design features in the design phase while considering their impact on safety functions. The methodology needs to show a clear result by identifying how the assessment objectives were met and how the screening process eliminated security design features from further consideration.

To assess these features for their impact on the PPS performance, each candidate security design feature should be evaluated against the limiting overall scenarios associated with each standard DBT scenario. Consideration should then be given to the improvement in the overall scenario margin or overall system effectiveness that can be achieved from a given improvement. This performance improvement could then be compared to the impact of the security design feature on the plant's safety functions.

2.5.2 Analyze Scenarios to Ensure Adversary Actions are within DBT Capabilities and Credible

The most transparent and logical way of ensuring that adversary actions are credible and within the DBT capabilities is to use the DBT, its accompanying guidance, and those references provided in the acceptable for use engineering publications listed in Appendix A. For any assumptions used in the creation of the adversary timeline that are not referenced directly to one of the acceptable engineering publications, a full description and justification should be provided. If an assumption is provided in the NRC publications and an applicant uses a different value, a sensitivity study using the NRC-provided value also should be included.

2.5.3 Analyze Scenario To Ensure Barrier Delay Times and Protective Force Actions are Credible

The most transparent and logical means of ensuring that barrier delay times and protective force actions are credible is to use those provided in the acceptable engineering publications listed in Appendix A. For any assumptions used in the creation of the protective force timeline that are not referenced directly to one of the acceptable engineering publications, a full description and justification should be included as part of the security assessment. If an assumption is provided in the NRC publications and an applicant uses a different value, a sensitivity study using the NRC-provided value also should be included.

2.6 High Assurance Evaluation

Step 5 of the high assurance evaluation process is depicted in Figure 2-16.

High Assurance Evaluation Process

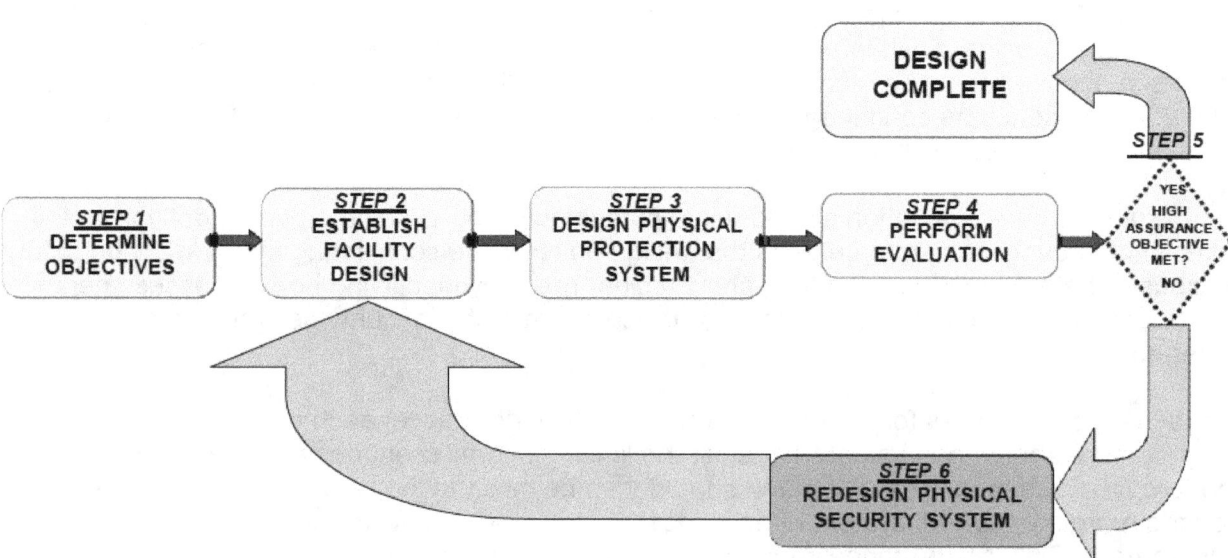

Step 5: High Assurance Evaluation

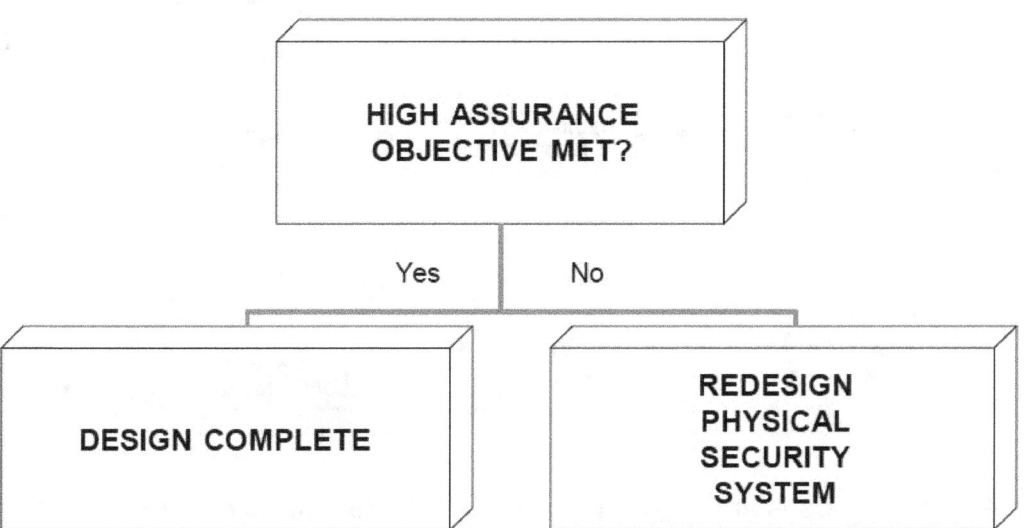

Figure 2-16 Security assessment process – step 5: was high assurance objective met?

The objective of a nuclear power plant's PPS is to provide high assurance of protection against the DBT of radiological sabotage. The performance-based measure for high assurance is satisfied if the physical protection system is capable of protecting all target sets to an acceptable level of risk of protective system failure. Quantitatively, this would be represented by a PPS with high overall system effectiveness. The applicant should clearly state the objective of high assurance performance and its bases used in the security assessment.

2.7   Redesign Physical Security System (Step 6)

Redesign of the PPS is Step 6 of the high assurance evaluation process as shown in Figure 2-17. If the PPS design does not meet the objective of high assurance as stated in Section 2.5, a redesign of the PPS should be performed. A redesign of the PPS causes an applicant to iterate through the process again, beginning at Step 3, in Section 2.4.

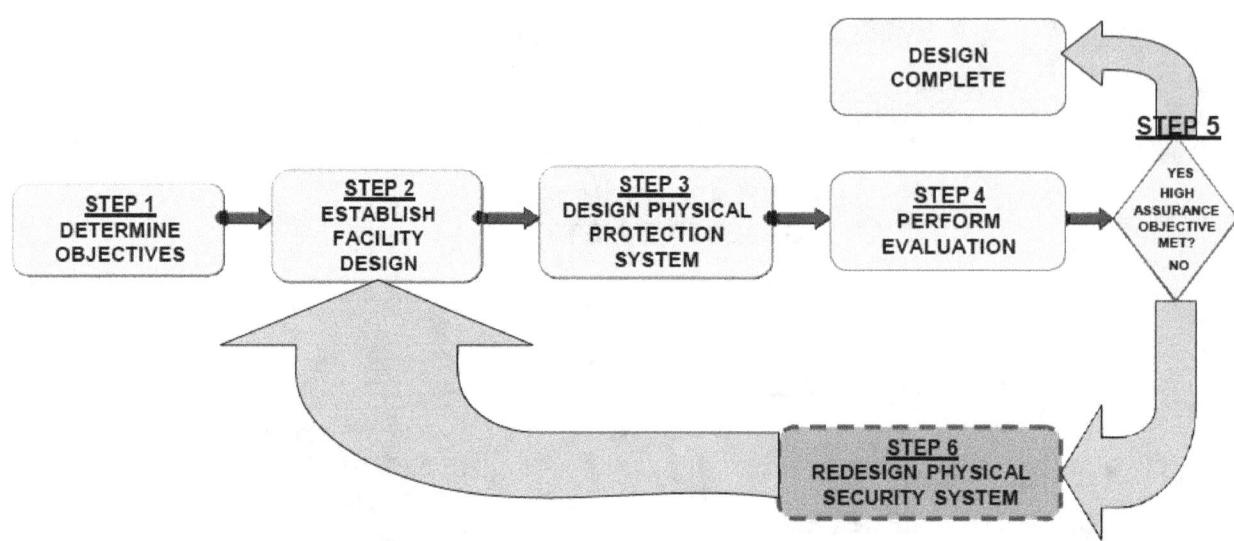

Figure 2-17 Security assessment process – step 6: redesign physical security system

Designing the PPS elements (Step 3) will not change the target sets in Step 2; instead, it will change the overall system effectiveness of protecting against the adversarial attack scenarios. It may also be possible to improve the physical protection system effectiveness by redesigning parts of the facility itself. If this is the case, the reiteration would begin at Step 2 in Section 2.3, since the facility design should be recharacterized and target sets should be re-screened to obtain the target sets applicable to that design.

After redesigning the PPS and modifying the facility design, the applicant would again undergo the evaluation process as outlined in Section 2.5 and test the design against the objective of high assurance described in Section 2.6. When the facility design and PPS elements meet the objective of high assurance, the evaluation process is complete. The design that meets the objective should be the design submitted to the NRC as part of the security assessment.

The iterative process presented above is designed to create a PPS that is efficient and effective at ensuring high assurance against the DBT. Through this method, applicants will gain insights about how to modify the PPS to better manage threats in the new post-September 11, 2001, environment. Insights gained during the iterations are also valuable and should be documented for submittal.

3. FORMAT AND CONTENT GUIDANCE

This section establishes the format and content guidelines for the security assessment. As described in Section 1.3, the format presented represents one that is acceptable to the NRC staff. Other formats could be acceptable if they provide an adequate basis for the findings. The level of detail needed in the documentation should be sufficient to enable the reader to understand and determine the validity of all input data and calculation models used, to enable the reader to understand the sensitivity of the results to key aspects of the physical protection system (PPS) including key analysis assumptions (i.e., identification of critical assumptions for which small changes could significantly impact the overall effectiveness of the PPS). The design information provided in the security assessment should reflect the most advanced state of the reactor facility design at the time of submission. It is not necessary to submit all the documentation generated while performing a security assessment for an NRC review, but basis documents, calculations, guidance and references should be cited and available in a clear and methodical format.

The following subsections outline the major parts of the security assessment. Note that appendices can be used for supplemental information that may contain more detailed data and diagrams that accompany the main body of the security assessment document.

Table 3-1 Security Assessment Table of Contents

Section		Title
1.0		**Executive Summary**
2.0		**High Assurance Evaluation**
2.1		Introduction
2.2		Purpose and Objectives
2.3		Scope and Facility Design
	2.3.1	Scope/Conduct of the Analysis
	2.3.2	Facility Characterization
	2.3.3	Security Assessment Parameters
2.4		Target Set Analysis
	2.4.1	Methods
	2.4.2	Results
2.5		Physical Protection System
	2.5.1	Iterative PPS Design Process
	2.5.2	Final PPS Design
	2.5.2.1	Detection Elements
	2.5.2.2	Delay Elements
	2.5.2.3	Response Elements
	2.5.2.4	Communication Elements
	2.5.3	Safety/Security Interface

Table 3-1 Security Assessment Table of Contents

Section	Title
2.6	Evaluation Methods and Results
2.6.1	Scenario Identification
2.6.2	Adversary Timeline Results
2.6.3	Protective Force Timeline
2.6.4	Evaluation Results
2.6.4.1	Overall System Effectiveness
2.6.4.2	Risk-informed Evaluation of Candidate Security Features
2.6.4.3	Sensitivity Studies
2.7	Discussion and Conclusions
3.0	**References**
Appendix	
A	Glossary/Abbreviations
B, C, etc.	Supplemental/Supporting Information

These sections include:

Section 1.0   Executive Summary

The executive summary should contain a brief overview of the security assessment process. Key results, findings, and insights should be introduced in this section.

Section 2.0   High Assurance Evaluation

Section 2.1   Introduction

The introduction should detail more information about the facility design being assessed and the assessment methodology. It should provide an overview of the methodology used to perform the security assessment and to evaluate the PPS effectiveness. This section should include a concise description of the major tasks in the methodology and how these interact with each other to generate the results of the capability of the PPS to protect the plant against the threats considered.

Section 2.2   Purpose and Objectives

The purpose of the security assessment should be described in this section. It should include information about the design stage of the reactor facility for which the assessment is being performed (e.g., construction permit, operating license, standard design approval, design certification, manufacturing license, or combined license). The scope of the security assessment will differ based on the particular stage of the application process for each reactor facility. For example, the security assessment for COL applicants may have a more comprehensive level of detail and include site-specific information, as compared to a security assessment for a design certification applicant.

Objectives should be reiterated as described in 10 CFR 73.55 with respect to prevention of significant core damage, sabotage of spent fuel, and theft and diversion of special nuclear materials. This section should also clearly identify the specific performance criteria the applicant used for the determination of high assurance, as discussed in Section 2.2.4 of the security assessment process. These criteria should allow for the determination of whether the PPS is capable of protecting all target sets with high assurance. The ability of the PPS to meet the applicant-defined objective of high assurance will be demonstrated in Section 7.0 of the security assessment.

While the NRC determines and provides the objectives of the security assessment, this section should be used to acknowledge receipt of the external and regulatory-driven information highlighted in Section 2.1 of the high assurance evaluation. This acknowledgment should include the titles and dates of such documents: 1) Regulatory Guide (RG) 5.69, "Guidance for the Application of Radiological Sabotage Design-Basis Threat in the Design, Development, and Implementation of a Physical Security Protection Program that Meets 10 CFR 73.55 Requirements," 2) Standard Set of Scenarios (Ref. 3), and 3) security engineering publications acceptable for use. At the onset of performing the assessment, verify with the NRC staff that these are the most current documents to use.

Section 2.3   Scope and Facility Design

Section 2.3.1 Scope and Conduct of the Analysis

This section should describe how the entire analysis has been conducted, including the quality assurance program, qualifications of the analysis staff (one paragraph resume for each assessment team member), any independent reviews performed and the protective measures used for any sensitive documentation (e.g., safeguards information) reviewed during the analysis.

Discussion of the scope of the assessment should include pertinent information about the applicant or licensee, the facility design or site being assessed, and any limitations that were placed on the assessment. Refer to Section 2.3.1 of this document for guidance on the scope of the security assessment for different stages of the application process. Examples of limitations include any limits imposed by the evaluation methods or model used (e.g., JCATS lacks the ability to model several elevations of a building simultaneously) or the skill or abilities of team members chosen to perform the assessment.

The security assessment should describe applicant staff participation (organization and role) and the extent to which the staff was involved in all aspects of the security assessment program. The security assessment should also contain a description of the peer review(s) performed, the result of the review team's evaluation, and a list of the review team members. Finally, the validation process of the input data for the security assessment should be described in the security assessment. The applicant should include a list of resources used to supplement the information necessary to perform a proper evaluation. For example, if an alternate source is used as opposed to one of the provided engineering publications listed in Appendix A, this should be cited and validated in this section.

Section 2.3.2 Facility Characterization

This section should summarize the results of the facility and site characterization for the design being submitted as part of the security assessment. This summary should focus on facility and site characteristics used in the target set analysis. It should include, but is not limited to, relevant facility drawings important to the security assessment (including buildings, room locations, etc.), important operational data, operational and maintenance configurations, the physical and environmental setting of the facility (site and property boundaries, adjacent facilities, etc.), access control points for normal and outage operating modes, types and numbers of employees and response time and capabilities of local law enforcement.

Include a top view drawing ("D" size) of the site that depicts physical security characteristics such as owner-controlled area boundaries, PA (protected area) boundaries, points of intrusion detection, VBS (vehicle barrier system), CAS (central alarm station), SAS (secondary alarm station), sally ports, vehicle checkpoints, delay features, active denial and delay features, hardened posts, delay barriers, fields of fire, anticipated fields of view for assessment devices, etc.

For design certification applicants, certain physical protection characteristics may be used so that a more thorough assessment can be performed. These characteristics include locating the protected area perimeter a minimum distance from vital areas, assuming the vehicle barrier systems are at the required standoff distance (RSD) (or minimum safe standoff distance) (see NUREG/CR-6190 for guidance) and identifying a number of armed responders (as a starting point for the design certification stage only). This section should therefore detail the RSD and the method used to calculate or identify it for the facility.

The design information provided in the security assessment should reflect the most advanced state of the design at the time of submission. Additionally, if the facility design changed as a result of the iterative security assessment process, any insights gained while iterating through the process that are directly related to the facility design also should be included in this section.

Section 2.3.3 Security Assessment Parameters

This section should include any security assessment parameters used in the security assessment to assess the PPS. Also included in this section should be features within the scope of the assessment being performed but its design is deferred to a future applicant that would have additional information to improve the security design feature. Ultimately, any security design issue identified by an assessment, but not addressed by a security design feature at any application stage would be identified by a security assessment parameter and should be addressed during the development of the security operational programs under the provisions of 10 CFR Part 73, "Physical Protection of Plants and Materials."

Section 2.4   Target Set Analysis

In this section, the assets that have been identified as targets of an adversary attack should be described. See RG 5.81, "Target Set Identification and Development for Nuclear Power Reactors," for detailed guidance for this section.

Section 2.4.1 Methods

Describe procedure(s) for the development and identification of target elements, and the analyses and methodologies used to determine and group the elements into target sets. These descriptions and their associated procedures should include, but are not limited to:

• 	process of target identification

- methodologies used to determine and group the target set elements

- screening criteria for achievable targets

- characterization and screening process used for identification of target sets

- if applicable, description and procedures used in alternative approaches

- target set analysis team qualification

- a listing of target set analysis input documents such as site layout drawings, PRA analyses, table-top analyses, etc.

- the process for considering the effects of cyber attacks upon individual or groups of target elements in each target set

- a listing of screened target sets, and achievable targets, and the associated bases for screening

Section 2.4.2 Results

Provide the most current set of achievable targets, as well as a list or description of those targets that were considered not achievable.

Verify that no single act, as bounded by the DBT, can disable an entire target set. Use blast affects references such as NUREG/CR-6190 and Regulatory Issue Summary 2003-06, "High Security Protected and Vital Area Barrier/Equipment Penetration Manual," dated March 20, 2003.

For each target set; a unique number should be assigned and the security assessment should include a table with the following information (Ref. 15):

- Target set number.

- Target set objective: a unique title for the target set that describes the overall general objective.

- Initiating event: for each target set, identify the malevolent act initiating event or events.

- Target set equipment: for each target set, the list of SSCs and operator actions that, if all are prevented from performing their safety function or prevented from being accomplished, will likely result in radiological sabotage, theft of special nuclear material or allow offsite release.

- Targeted equipment locations: for each target set, identify the locations of the SSCs and operator actions that are identified as equipment for that particular target set.

- Adversary actions: describe the objectives and actions, in general terms that the adversary force would need to complete to achieve significant core damage, spent fuel sabotage or theft and diversion of radioactive materials.

- Target resiliency: categorization of target resiliency and bases (based upon resilience to DBT attack).[7]

- Credited operator actions: consider and list appropriate preventative equipment and operator actions in the target set. If operator actions are credited, then provide a listing of the six required provisions (see RG 5.81 for details).

- Estimated time to core damage/spent fuel sabotage: the estimated time after all targets that make up a target set to achieving significant core damage or spent fuel sabotage.

- Anticipated results/basis: a brief description of the anticipated outcome and the basis for that outcome with regard to why significant core damage, spent fuel sabotage, or theft of nuclear materials will occur. The anticipated results should occur within a short enough time period, such that effective operator mitigation is prevented.

- Likelihood of exceeding Part 100: determine the potential 10 CFR Part 100 (Ref. 28) exposure impact or need for protective action recommendations.

- Additional considerations: a brief discussion with regard to the basis for potentially achieving mitigation as a result of operator actions (after a target set has been lost). Additionally, any other noteworthy comments should be included here.

Section 2.5   Physical Protection System

Section 2.5.1 Iterative PPS Design Process

The applicant should provide information about the final combination of PPS elements. If multiple iterations were necessary through the process, reporting the intermediate PPS elements that were considered during these iterations is unnecessary. It is recommended, however, that these intermediate assessments be held in onsite documentation. Provide a summary of the major insights gained during the iterative evaluation process that are related to the addition of PPS elements.

Section 2.5.2 Incorporation of Security Design Features

The applicant should provide a description of the physical protection systems for the plant design, including the people, procedures, and the detection, delay, and response characteristics proposed for the protection of assets or facilities against theft, radiological sabotage, or other malevolent human attacks. This section should detail how and where the security design features, identified as a result of the security assessment, have been integrated into the design. The applicant should list the security functions for the plant. This listing should include the detection, delay, and response elements of the PPS and systems, structures, and components with their associated security functions (if applicable) for the stage of design being evaluated. Also included should be an explanation of how each feature provides or enhances the capability of the facility to protect the targets sets and related elements against an adversary possessing the DBT characteristics.

---

[7]   See definition of target resiliency in Appendix A of this report.

Additional guidance is provided below for each PPS element. The description for these elements should include, but is not limited to, the following areas:

Section 2.5.2.1 Detection Elements

A list of intrusion sensors (internal and external) used in the final PPS design should be provided. Consider including humans as detection elements (where appropriate and in accordance with 10 CFR 73.55(e)(4)). Describe how interactions between sensor hardware and the physical environment were considered and reconciled. Additionally, list the alarm assessment subsystems used in the PPS design. If human alarm assessment was used in place of video alarm assessment subsystems, provide an explanation for this choice, and detail the availability, reliability, environmental, and communications requirements. Describe the entry control subsystems used in the final PPS design. This description should identify a distinction if the subsystems are for personnel or vehicle control and whether they are manual, machine-aided manual, or automatic. Finally, discuss the alarm communication and display subsystem (which transports sensor alarm and video information to a central location and presents the information to a human operator).

Details should be provided about the placement and protection of the CAS and SAS. These should be designed within the PPS so that they have functional equivalent capabilities such that no single act can disable the function of both the CAS and SAS. If applicable, the design can be verified using blast effects guidance such as NUREG-6190, RIS 2003-06, "High Security Protected and Vital Area Barrier/Equipment Penetration Manual," dated March 20, 2003, and RIS 2005-09, "High Security Protected and Vital Area Barrier Breaching Analysis," dated June 6, 2005 (Ref. 29).

Section 2.5.2.2 Delay Elements

List the passive and active barriers used in the final PPS design for access delay. Include passive delays such as fences, gates, vehicle barriers, walls, floors, roofs, doors, windows, grilles, utility ports, and other elements. Include active delays such as nonlethal weapons, dispensable materials, and deployable barriers. Note that delay elements located before the first detection point on the adversary pathway are not included because the scenario timelines start at first detection.

Section 2.5.2.3 Response Elements

Discuss response capability and strategy used in the final PPS design. Include both immediate and delayed response capability. Additionally, describe how response communication was integrated as a part of the response strategies. This section should also include active denial systems (e.g., remotely operated weapons, munitions based active denial systems).

Section 2.5.2.4 Communication Elements

Describe the communication systems related to security that are used for both onsite and offsite communications. Describe duress alarms and methods of secondary onsite communication between the response force personnel and the CAS and SAS.

Section 2.5.3 Inclusion of Security Design Features in Plans and Appendices

This section should indicate where security design features, identified as a result of the security assessment, are delineated in security plans and appendices required by 10 CFR 73.55. Design certification applicants should identify those features that are to be included in security plans by future applicants that reference their design.

Section 2.5.4 Safety/Security Interface

The applicant should demonstrate that the interface between safety and security was considered when designing PPS elements and that there are no actual or potential interactions in which safety design or operational (including maintenance) activities may adversely affect security activities or vice versa. This section should also include a description of the process used to demonstrate that the safety and security interactions were appropriately addressed. Specific discussion should be included on the impact and treatment of plant operational modes, PPS maintenance, and maintenance for equipment that is included as elements within the target sets.

Section 2.6   Evaluation Methods and Results

This section addresses the way the facility design was evaluated and shown able to achieve the high assurance objective of overall physical protection system effectiveness.

Section 2.6.1 Scenario Identification

For each of the standard scenarios provided by the NRC, provide a description of the method used to identify the set of overall scenarios that will be used to determine the effectiveness of the PPS. Information should include how potential adversary pathways were identified, including consideration given to target access points, detection devices, traversal distances, protective features, and anticipated protection force routes. For each scenario, the credited detection devices should be clearly identified and include the detection device(s) that is/are used to establish the point of detection in the scenario's timeline (e.g., an accumulation of detection probability acquired along an adversary pathway may identify the point of detection in a scenario timeline) (see Section 2.5.1.5.). Additionally, any considerations used to determine the most vulnerable pathways should be described in this section. Similar descriptions should be included to explain how the protective force pathways were determined. These descriptions should include the types of security response and how multiple security responses are assessed (e.g., using a bounding response).

Section 2.6.2 Adversary Timeline Results

Submit the depictions (e.g., logic diagrams, event trees) of the adversary timelines with the lowest margins for each standard DBT scenario that the NRC provided. For any assumptions used in creation of the adversary timeline that are not referenced directly to one of the acceptable engineering publications listed in Appendix A, full justification should be included. If an assumption is provided in the NRC publications and an applicant uses a different value, a sensitivity study using the NRC-provided value also should be included.

Section 2.6.3 Protective Force Timeline

Provide a description of the method used to assess the protective force timeline for each standard DBT scenario that the NRC provided. This description should include the location of each of the CIPs that the applicant has identified. For an existing facility, the method used to assess the protective force timeline should be actual protective force response times documented by performance tests. The average or mean times from the performance tests, including one standard deviation, should be used. For assaults other than land-based attacks, the minimum safe standoff (or required) distance for protection against vehicle-borne improvised explosive devices also should be identified. Additionally, a description and a depiction of the milestones in the timeline should also be provided, including the point of communication, the point in which the CIP is reached, and the point in which weapons are readied. Finally, the time elapsed between each of these milestones also should be reported. For any assumptions used in creation of the protective force timeline that are not referenced directly to one of the acceptable engineering publications listed in Appendix B, full justification should be included. If an assumption is provided in the NRC publications and an applicant uses a different value, a sensitivity study using the NRC-provided value also should be included.

Section 2.6.4 Evaluation Results

This section should explain how overall system effectiveness was determined. Describe the analysis method used for the evaluation. If table-top reviews are used, detailed descriptions of the methodologies are recommended. If other analysis tools, such as a JCATS model, are used as a modeling and simulation tool, all assumptions (input variables) used in the model (e.g., Ph and Pk data) and their sources should be detailed. For any assumptions used in the evaluation of overall system effectiveness that are not referenced directly to one of the acceptable engineering publications listed in Appendix B, full justification should be included. If an assumption is provided in the NRC approved publications and an applicant uses a different value, a sensitivity study using the NRC referenced value also should be included. When making reference to a simulation model list the model revision number.

Section 2.6.4.1 Overall System Effectiveness

Overall system effectiveness of the PPS can be reported qualitatively and quantitatively. It should be demonstrated for an overall scenario, defined as a unique combination of target set, entry point, exit point (for theft and diversion), protective force response, and standard DBT scenario. Yet, it should be noted that overall system effectiveness does not need to be demonstrated for each individual target, as long as at least one target in each target set is shown to remain protected with high assurance.

The results in this section should include a list of the most advantageous to an adversary overall scenario, specifically describing those overall scenarios that have the lowest overall system effectiveness values or with timelines demonstrating the least margin. This set should include the worst overall scenario corresponding to each standard scenario. If the applicant chooses to combine or bound overall scenarios (i.e., using the most advantageous adversary pathway for a standard scenario) to simplify the reporting process, adequate bases and description of the approach should be included.

The qualitative results should be provided using an integrated adversary and protective force timeline, with a shifted protective force timeline beginning at the CDP. The results for each of the worst overall scenarios, as well as an integrated timeline, should be depicted with a CDP and a CIP. Additionally, adequate margin (time from the point of detection to the CDP) to enter into a defensible position and ready weapons should be demonstrated. If the security assessment uses average or mean times for the adversary and responder timelines, an adequate margin on the order of one standard deviation should be included in the timeline assessment to demonstrate meeting the objective of high assurance.

Quantitatively, milestones in the integrated timeline for each of the worst overall scenarios should be assigned probabilities. These values should be provided in a tabular format, such as the example provided in Table 2-3, and should include, but are not limited to, the P_D, P_I, and P_N and the bases by which these probabilities were calculated. As the product of these probabilities is the overall system effectiveness for a scenario, this value (P_E) also should be provided. The NRC-recommended probability of detection at the protected area boundary is 90 percent detection rate with 95 percent confidence. If this is not met, an explanation is recommended as to why a lesser percentage is acceptable to achieve overall system effectiveness.

Identification of technical knowledge gaps responsible for significant uncertainties in scenarios timelines and quantitative measure values also should be identified in this section.

Section 2.6.4.2 Risk-Informed Evaluation of Candidate Design Features

This section addresses the risk-informed assessment of candidate security design features. It should include the process used to select and assess the candidate design features included in the security assessment. The section also should list the features evaluated and the assessment results.

Section 2.6.4.3 Sensitivity Studies

This section should be used to highlight any sensitivity studies performed on the assumptions used in the evaluation. Sensitivity studies should be performed when using assumptions not provided in the NRC engineering publications listed in Appendix B.

Section 2.7   Discussion and Conclusion

This section should summarize the high assurance evaluation process and results, including a brief discussion of how the security assessment demonstrates that the proposed design, operation, and maintenance of the facility meets the requirements established by 10 CFR 73.55. Additionally, any insights gained from sensitivity studies can be included in this section.

Any insights gained during the high assurance process, potentially including those from the iterative design process, should be described here. The high assurance evaluation process implemented by the applicant, if iterative, would produce an initial evaluation on the existing facility design and PPS elements, some number of intermediate (trial and error) evaluations, in which facility design or PPS elements are added to the design, and a final evaluation, in which the objective of high assurance is met. If the applicant-performed iterative runs through the process, this section should briefly describe any insights gained from these evaluations. This section should include the methodology the applicant used for the modification and addition of the facility design and PPS elements during the process.

Section 3.0   References

This section should list all documents referenced in the security assessment. References should be listed in a methodical fashion (either alphabetically or numbered by chronological order of reference in the security assessment). The analysis methodology used in the assessment should be referenced in this section.

Appendix A   Glossary/Abbreviation

An appendix should establish a list of all abbreviations used in the security assessment and any key words that may be appropriate to define in the glossary.

Appendix (Additional)

The applicant should use additional appendices for supplemental information, drawings, diagrams, data tables, calculations (e.g., blast analyses), etc. that are not directly necessary in the security assessment, but may be pertinent to reproducing calculations or validating information.

4. REFERENCES

1.   Title 10 of the *Code of Federal Regulations* (10 CFR) 73.55, "Requirements for Physical Protection of Licensed Activities in Nuclear Power Reactors against Radiological Sabotage."

2.   Title 10 of the *Code of Federal Regulations* (10 CFR) Part 73, "Physical Protection of Plants and Materials."

3.   U.S. Nuclear Regulatory Commission (NRC). "Nuclear Power Plants Security Assessment Standard Set of Scenarios," December 20, 2007.

4.   U.S. Nuclear Regulatory Commission (NRC). "Policy Statement on Regulation of Advanced Reactors," 73 FR 60612; October 14, 2008.

5.   SECY-05-120, "Security Design Expectations for New Reactor Licensing Activities," July 6, 2005, and SRM to SECY-05-120, "Staff Requirements - SECY-05-120 - Security Design Expectations for New Reactor Licensing Activities," September 9, 2005.

6.   SRM to SECY-06-0204, "Staff Requirements - SECY-06-0204 - Proposed Rulemaking - Security Assessment Requirements for New Nuclear Power Reactor Designs," April 24, 2007 (ADAMS Accession No. ML070520692).

7.   SAND2007-5591, "Nuclear Power Plant Security Assessment Technical Manual," Sandia National Laboratories, Albuquerque, NM, September 2007. http://www.sandia.gov (ADAMS Accession No. ML072620172).

8.   Title 10 of the *Code of Federal Regulations* (10 CFR) Part 50, "Domestic Licensing of Production and Utilization Facilities," Appendix B, "Quality Assurance Criteria for Nuclear Power Plants and Fuel Reprocessing Plants."

9.   U.S. Nuclear Regulatory Commission (NRC). Regulatory Guide 5.81, "Target Set Identification and Development for Nuclear Power Reactors." **(Includes security-related or safeguards information and is not publicly available.)**

10.   "The Design and Evaluation of Physical Protection Systems," 2001, and "Vulnerability Assessment of Physical Protection Systems," 2006, Garcia, Mary Lynn, Sandia National Laboratories, http://www.sandia.gov, published by Elsevier Butterworth-Heinemann, Burlington, MA (were used as a basis for the development of Section 2.0 of this guide, "High Assurance Evaluation Guidance.")

11.   Title 10 of the *Code of Federal Regulations* (10 CFR) 73.1, "Purpose and Scope."

12.   U.S. Nuclear Regulatory Commission (NRC). Regulatory Guide 5.69, "Guidance for the Application of Radiological Sabotage Design Basis Threat in the Design, Development, and Implementation of a Physical Security Protection Program that Meets 10 CFR 73.55 Requirements." **(Includes security-related or safeguards information and is not publicly available.)**

13. NUREG/CR-6190, Revision 1, "Protection Against Malevolent Use of Vehicles at Nuclear Power Plants," March 17, 2004, U.S. Army Corps of Engineers, Omaha, NE. **(Includes security-related or safeguards information and is not publicly available.)**

14. NUREG-0800, "Standard Review Plan for the Review of Safety Analysis Reports for Nuclear Power Plants (LWR Edition)," March 2007 (ADAMS Accession No. ML070660036).

15. SECY-03-0052, "Staff Recommendations for Revisions to the Design Basis Threat Statements (U)," April 7, 2003. **(Includes security-related or safeguards information and is not publicly available.)**

16. Title 10 of the *Code of Federal Regulations* (10 CFR) 73.2, "Definitions."

17. U.S. Nuclear Regulatory Commission (NRC). Regulatory Guide 5.76, "Physical Protection Programs at Nuclear Power Reactors." **(Includes security-related or safeguards information and is not publicly available.)**

18. "A Method to Assess the Vulnerability of U.S. Chemical Facilities," U.S. Department of Justice, Office of Justice Programs, 2002, http://www.justice.gov.

19. NUREG-1959, "Intrusion Detection Systems and Subsystems: Technical Information for NRC Licensees," March 2011 (ADAMS Accession No. ML11112A009).

20. SAND2001-2168, "Technology Transfer Manual—Access Delay Technology, Volume 1," Sandia National Laboratory, Albuquerque, NM, http://www.sandia.gov.

21. Regulatory Issue Summary (RIS) 2003-06, "High Security Protected and Vital Area Barrier/Equipment Penetration Manual," U.S. Nuclear Regulatory Commission, Washington, DC, March 20, 2003. **(Includes security-related or safeguards information and is not publicly available.)**

22. Nuclear Energy Institute 09-05 "Guidance on the Protection of Unattended Openings that Intersect a Security Boundary."

23. Title 10 of the *Code of Federal Regulations* (10 CFR) 73.58, "Safety/Security Interface Requirements for Nuclear Power Reactors."

24. U.S. Nuclear Regulatory Commission (NRC). Regulatory Guide 5.74, "Managing the Safety/Security Interface," Revision 0, June 2009 (ADAMS Accession No. ML091690036).

25. "Joint Conflict and Tactical Simulation (JCATS)," United States Joint Forces Command, 2006, http://www.jfcom.mil.

26. U.S. Nuclear Regulatory Commission (NRC). Regulatory Guide 5.68, "Protection Against Malevolent Use of Vehicles at Nuclear Power Plants" (ADAMS Accession No. ML003739379).

27. International Atomic Energy Agency, Engineering Safety Aspects of the Protection of Nuclear Facilities Against Sabotage, Nuclear Security Series No. 4, Vienna, Austria, January 2007, http://www.iaea.org.

28.   Title 10 of the *Code of Federal Regulations* (10 CFR) Part 100, "Reactor Site Criteria."

29.   Regulatory Issue Summary (RIS) 2005-09, "High Security Protected and Vital Area Barrier Breaching Analysis," June 6, 2005. **(Includes security-related or safeguards information and is not publicly available.)**

30.   ASME/ANS RA-Sa-2009, "Addenda to ASME/ANS RA-S-2008 Standard for Level 1/Large Early Release Frequency Probabilistic Risk Assessment for Nuclear Power Plant Applications," 2009, http://www.asme.org.

31.   Title 10 of the Code of Federal Regulations (10 CFR) Part 73, "Physical Protection of Plants and Materials," Appendix C, "Nuclear Power Plant Safeguards Contingency Plans."

32.   Title 10 of the *Code of Federal Regulations* (10 CFR) 100.21, "Non-seismic Site Criteria."

Appendix A

Glossary

APPENDIX A GLOSSARY

Achievable Target Element   A target element that is within the capabilities included in the design-basis threat.

Adversary Timeline   An assessment of the impact of the physical protection functions of detection, delay, and response on the adversary for a given DBT scenario, given specific entry points and a given target set.

Critical Detection Point   The point on the path where path delay just exceeds protective force arrival time. This point is found by starting at the end of the adversary path, and adding up path delays until this value just exceeds protective force time (Ref. 10).

Critical Interruption Point   A location that maximizes the tactical and strategic capabilities for the response to interrupt and successfully neutralize most threats considered in the DBT (Ref. 27).

Design-Basis Threat (DBT)   A description of all the attributes and characteristics of the threat, including the type of adversary and the tactics and capabilities associated with the threat, provided by NRC.

DBT Scenario   A description of a specific set of the attributes and characteristics of the DBT provided by the NRC, which may include the number of adversaries, the type of weapons or tools they would use, tactics, and the number and type of entry points.

Event Tree   A logical diagram that begins with an initiating event or condition and progresses through a series of branches that represent expected system or operator performance that either succeeds or fails and arrives at either a successful or failed end state (ASME 2005) (Ref. 30).

Initiating Event   Any event, either internal or external to the plant, that perturbs the steady state operation of the plant (if operating), thereby initiating an abnormal event such as a transient or loss of coolant accident (LOCA) within the plant (Ref. 30).

Interruption   Arrival of responders at a deployed location to halt adversary progress or activation of engineered delay or denial systems at a deployed location to halt adversary progress.

Margin   The time elapsed on the integrated adversary and protective force timeline between the point of detection and the critical detection point or the time elapsed, after the response team has entered protective positions with weapons at the ready, and engagement of the adversary.

Neutralization	The defeat of the adversaries by responders through the use of a small arms conflict or remote engagement with automated weaponry, or certain containment strategies.
Overall Scenario	A unique combination of a DBT scenario, adversary entry point, adversary exit point (for theft and diversion), protective force response and target set.
Overall System Effectiveness	A probabilistic calculation of the effectiveness of the physical protection system (PPS) to detect the adversary, delay the adversary such that responders can intercept the adversary, ideally by reaching their protective positions, and neutralize the adversary.
Physical Protection Systems	The integration of people, procedures, and equipment for the protection of assets or facilities against theft, radiological sabotage, or other malevolent human attacks (Ref. 10).
Protective Force Timeline	An assessment of the time after initial detection of adversarial activity it will take for one or more members of the security force to reach a location where an adversary's path can be interrupted.
Radiological Sabotage	Any deliberate act directed against a plant or transport in which an activity licensed pursuant to the regulations in 10 CFR Part 73, "Physical Protection of Plants and Materials," is conducted, or against a component of such a plant or transport, which could directly or indirectly endanger the public health and safety by exposure to radiation (10 CFR 73.2) (Ref. 16).
Safeguards Contingency Plan	A documented plan to give guidance to licensee personnel to accomplish specific defined objectives in the event of threats, thefts, or radiological sabotage relating to special nuclear material or nuclear facilities licensed under the Atomic Energy Act of 1954, as amended. (Appendix C, "Nuclear Power Plant Safeguards Contingency Plans," to 10 CFR Part 73 (Ref. 31.)
Security Assessment	An evaluation of the reactor facility design, which: 1) identifies target sets and, for selected scenarios, performs a systematic evaluation using risk evaluation methodologies that demonstrate the ability of the design to meet the performance objectives of 10 CFR 73.55(a), 2) identifies security design features to be incorporated into the design of the reactor facility, which indicate that security functions can be accomplished, to the maximum extent practical, without undue reliance upon operational security programs that are required as a part of the security plans under 10 CFR 73.55, "Requirements for Physical Protection of Licensed Activities in Nuclear Power Reactors Against Radiological Sabotage," and 3) demonstrates that the design features and operational recovery actions incorporated into the nuclear power plant and programs provide for mitigation of the effects of an

attack resulting in a loss of large areas of the facility because of explosions or fires, in accordance with 10 CFR 50.54(hh).

Security Assessment Parameters

The characteristics of parameters of a site where the nuclear power plant or reactor is to, or may, be used either as postulated in the security assessment or as identified in accordance with 10 CFR 100.21(f) (Ref. 32); security design features which are outside the scope of the design being addressed at the particular stage of the regulatory process, which are postulated in a security assessment, and features of a physical security program under 10 CFR 73.55, which are postulated in a security assessment.

Security Design Features

The structures, systems, and components of a nuclear power plant and their layout that are relied upon to either detect, delay, or respond to an attack against target sets of a nuclear power plant by an adversary possessing the characteristics of the DBT

Security Functions

Those functions necessary to: detect, delay, or respond to an attack against target sets of a nuclear power plant by an adversary possessing the characteristics of the DBT or provide conditions before, during, and after a malevolent event, that facilitate actions to occur that mitigate the effects of circumstances associated with a loss of large areas of the facility of explosions or fires.

Significant Core Damage

Non-incipient, non-localized fuel melting and/or core disruption (10 CFR 73.2).

Target Resiliency

Describes a target's robustness or resistance to a DBT attack. A resilient target could have characteristics that include multiple barriers, large distances from related targets or from the fence-line, and robust construction.

Target Set

The minimum combination of equipment or operator actions which, if all are prevented from performing their intended safety function or prevented from being accomplished, would likely result in significant core damage (e.g., nonincipient, nonlocalized fuel melting and/or core destruction) or a loss of spent fuel pool coolant inventory and exposure of spent fuel, barring extraordinary actions by plant operations.

Vital Area

Any area that contains vital equipment (10 CFR 73.2).

Vital Equipment

Any equipment, system, device, or material, the failure, destruction, or release of which could directly or indirectly endanger the public health and safety by exposure to radiation. Equipment or systems, which would be required to function to protect public health and safety following such failure, destruction, or release are also considered to be vital (10 CFR 73.2).

Appendix B

Security Engineering Publications Acceptable for Use

APPENDIX B SECURITY ENGINEERING PUBLICATIONS ACCEPTABLE FOR USE

SECURITY ENGINEERING REFERENCES ACCEPTABLE FOR USE IN THE DESIGN OF PHYSICAL PROTECTION SYSTEMS

Blast Effects

1. PDC-TR-01-01, Revision 1, "Structural Assessment of Spent Fuel Pools Attacked with a Sophisticated Sabotage Threat," U.S. Army Corps of Engineers, Omaha, NE, September 2006. Safeguards information.

2. PDC-TR-01-02, Revision 1, "Structural Assessment of Spent Fuel Pools Attacked with an Unsophisticated Sabotage Threat," U.S. Army Corps of Engineers, Omaha, NE, September 2006. Safeguards information.

3. Single Degree of Freedom Blast Design Spreadsheet (SBEDS) Version 4.1 Software and Methodology Manual, U.S. Army Corps of Engineers, Omaha, NE, March 13, 2009. Unclassified.

4. Regulatory Information Summary 2005-09, "High-Security Protected and Vital Area Barrier Breaching Analysis," U.S. Nuclear Regulatory Commission, Washington, DC, June 6, 2005. Safeguards information.

5. "Waterborne Sub-Surface Blast Effects to the Design-Basis Threat," D. Sulfredge, Oak Ridge National Laboratory, Oak Ridge, TN, November 10, 2003. Safeguards information.

6. "Guidance for Using Underwater Explosion (UNDEX) Data for Estimating Loads on Submerged Targets," D. Sulfredge, Oak Ridge National Laboratory, Oak Ridge, TN, and B. Tegeler, U.S. Nuclear Regulatory Commission, Washington, DC, November 2003. Unclassified.

7. FM 5-250, "Explosives and Demolitions," Department of the Army, Washington, DC, June 30, 1999. Restricted to government agencies and their contractors, export controlled.

8. Air Force Manual (AFMAN) 91-201, "Explosive Safety Standard," U.S. Air Force, Washington, DC, May 1, 1999. Unclassified.

9. DOETIC-11268, "Manual for the Prediction of Blast and Fragment Loading for Structures," U.S. Department of Energy, Washington, DC, July 1992. Unclassified.

10. Conventional Weapons Effects (CONWEP) Software and Manual, U.S. Army Corps of Engineers, Engineering Research and Development Center (ERDC), Vicksburg, MS, August 20, 1992. Restricted to government agencies and their contractors.

11. TM 5-1300, "Structures to Resist the Effects of Accidental Explosions," U.S. Department of Defense, Washington, DC, November 19, 1990. Unclassified. (Also designated as Air Force AFR 08-22 and Navy NAVFAC P-3897.)

12. Window Glazing Analysis Response and Design (WINGARD) Software, U.S. General Services Administration (GSA), Washington, DC. Restricted. (Available at www.oca.gsa.gov.)

Vehicle Barrier Systems/Blast Effects

13. NUREG/CR-6190, "Protection Against Malevolent Use of Vehicles at Nuclear Power Plants," U.S. Army Corps of Engineers, Omaha, NE, March 17, 2004. Safeguards information.

Vehicle Barrier Systems

14. Department of Defense and Department of State Certified Vehicle Barrier List (updated periodically, available at https://pdc.usace.army.mil/library/BarrierCertification/.) Unclassified.

15. SD-STD-02.01, "Certification Standard, Test Method for Vehicle Crash Testing of Perimeter Barriers and Gates," Revision A, Department of State, Washington, DC, March 2003. Unclassified.

16. NUREG/CR-4250, "Vehicle Barriers: Emphasis on Natural Features," U.S. Nuclear Regulatory Commission, Washington, DC, July 1985. Unclassified.

Detection, Delay, Communications, Security Systems, etc.

17. Regulatory Information Summary 2003-06, "High-Security Protected and Vital Area Barrier/ Equipment Penetration Manual," U.S. Nuclear Regulatory Commission, Washington, DC, March 20, 2003. Safeguards information.

18. SAND-2001-2168, "Technology Transfer Manual, Access Delay, Volume 1," Sandia National Laboratories, Albuquerque, NM, August 2001. In addition, the entire Technology Transfer Manual Series: SAND99-2390, SAND-2000-2142, SAND2004-2815P, SAND99-391, SAND99-2388, SAND99-2392, and SAND99-2389. Unclassified controlled nuclear information.

Ballistics

19. UL 752, "Standard for Bullet-Resisting Equipment," Underwriters' Laboratories, Northbrook, IL, December 21, 2006. Unclassified.

20. NIJ Standard 0108.01, "Ballistic-Resistant Protective Materials," National Institute of Justice, Washington, DC, September 1985. Unclassified.

21. ASTM F2656-07, "Standard Test Method for Vehicle Crash Testing of Perimeter Barriers," American Society for Testing and Materials International, West Conshohocken, PA, 2007. Unclassified.

Appendix C

Security Assessment Modeling Tools

APPENDIX C SECURITY ASSESSMENT MODELING TOOLS[8]

Security Assessment Modeling Tools

This appendix identifies some of the tools acceptable for use as part of the security assessment process. While these tools are identified as acceptable for use, from a performance-based perspective, the applicant may use other methods that are demonstrated to be equally effective. See the "Nuclear Power Plant Security Assessment Technical Manual," SAND2007-5591, for additional information on security assessment modeling tools.

1.0   Estimate of Adversary Sequence Interruption (EASI)

The EASI model is a quantitative method of evaluating the effectiveness of physical protection resources through segments and the physical protection systems (PPS) as a whole against the adversary's pathway. By evaluating detection and delay capabilities at each segment of the adversary's path and considering the probability of alarm communication and interpretation, and protective force deployment time, EASI measures the capability that the guard force will be able to interrupt the adversary before completing its goal.

The EASI evaluation method considers that the protective force must be notified and be strategically deployed while there is still sufficient time remaining in the adversary sequence. The EASI probabilistic evaluation sequence evaluates PPS effectiveness by accounting for the following features:

1.   The probability of detection at each sequence throughout the PPS based on inherent and adopted measures.

2.   The probability and time that the alarm will transmit and be interpreted accurately at the facility alarm station.

3.   The probability and time it takes for the alarm to be communicated from the facility alarm station to the PPS protective force.

4.   The time it takes for the protective force to deploy to their tactical positions.

The EASI software may be obtained at:
http://www.elsevierdirect.com/v2/companion.jsp?ISBN=9780750683524/.

2.0   Joint Conflict and Tactical Simulation (JCATS)

JCATS is a force-on-force computer-assisted simulation system developed to exercise commanders and their staff in the command and control of combined arms operations in urban terrain environments. It can model up to 10 parties with rules for reactive behaviors, allowing it to simulate realistic operations. It can compute the probability of the adversary reaching the target, as well as the number of protective and adversarial forces killed. JCATS can additionally

---

[8]   The assessment and modeling-type tools used throughout this NUREG and described in detail in this Appendix are for illustration purposes only. The NRC does not endorse any specific assessment or modeling-type tool for security assessments.

assist in defining the minimum number of protective force personnel and optimum response strategy necessary to achieve PPS effectiveness.

It should be noted that JCATS does not calculate a probability of interruption and does not include probabilities of detection in the facility model. It provides an estimate of the outcome of an engagement and if one runs enough engagements, it can aid in the derivation of an estimate for probability of neutralization (P_N).

3.0   Simajin

Simajin performs similar functions to JCATS, but addresses a number of limitations found in JCATS before the new post-September 11, 2001, threat environment (Grover, 2006). For example, Simajin can manage the increasing number of adversaries in the new DBT and the additional protective force required to counter these adversaries and their capabilities. Additionally, Simajin has more thorough data output capabilities, as well as the ability to receive more statistically valid data. It is capable of simulating force-on-force scenarios down to the single person level of detail.

While JCATS is considered acceptable for the purposes of calculating P_N, the NRC has not yet certified Simajin as acceptable. Another option the applicant can use to estimate a probability of neutralization is to use expert opinion.

4.0   Analytic System and Software for Evaluating Safeguards and Security (ASSESS)

The Analytic System and Software for Evaluating Safeguards and Security (ASSESS) software was developed for use by U.S Department of Energy sites to determine how effectively physical protection and material control and accountability systems protect against a spectrum of insider, outsider, and some collusion threats. (Effectiveness is measured as a probability, the probability of system effectiveness (P_E), which measures how likely it is that an attack by a certain type of adversary will be defeated given that it occurs.) The software addresses theft of special nuclear material and sabotage at these sites. The software was developed jointly by Sandia National Laboratories and Lawrence Livermore National Laboratories.

ASSESS represents protection around a target such as a weapon production facility, reactor facility, or weapon storage site in terms of concentric layers of delay and detection features (up to 10 layers are allowed). Detection features on a layer include intrusion sensors, protective force personnel, contraband checks, and access control features. Up to 15 elements (doors, sensor intrusion zones, walls, etc.) are allowed on each layer. ASSESS then determines the critical path for the adversary to take through the facility, which minimizes the probability of interruption, (P_I), the probability that the adversary will be detected while the security forces have enough time to respond.

ASSESS performance databases, developed based on U.S. national laboratory tests and expertise, describe the effectiveness of delay and detection features against a variety of adversary threats so the analyst does not need to develop all of these values. These default values can be overridden to reflect data from site-specific delay or sensor performance tests. The description of the critical path also tells the best way for an adversary to penetrate a barrier or sneak contraband past a screening portal; this description suggests performance tests that should be run at the different elements in the facility.

Some of the strengths of ASSESS include:

- It can search quickly through thousands or millions of paths, different operating conditions, and different threats and targets; combat simulations can at best evaluate about a dozen different combinations of intrusion paths, conditions, threats, and targets each day.

- It addresses insider and outsider threats, some of which will not attack using a frontal assault (for example, the insider).

- It determines whether potential delay and detection upgrades actually improve overall protection system effectiveness as opposed to just changing the critical path at the same performance level.

Some factors that ASSESS does not address:

- It does not allow for comparison of different response tactics, staffing levels, and weapons.

- It is an analytical model that does not include the natural randomness of battle, such as missed shots, confusion, fatigue, etc.

ASSESS is used to identify paths and conditions with low detection and delay that can then be evaluated in more detail with computer combat simulations or force-on-force exercises. ASSESS can also screen detection and delay upgrades to find those that actually improve system effectiveness. It includes a very simple model (i.e., it uses a semi-Markov[9] representation of a battle) meant to address basic response issues such as numbers of response units to employ and weapons to use, but it is very crude and doesn't allow the fidelity or detail of a combat simulation. This model was designed to provide a quick, consistent way to determine the P_N, the likelihood that the response can defeat or neutralize the adversaries, once they have been detected. ASSESS combines this P_N with P_I for the critical path to develop probability of system effectiveness $(P_E) = P_I * P_N$ to determine the effectiveness of detection, delay, and response.

5.0   Adversary Time Line Analysis System (ATLAS)

The Adversary Time Line Analysis System (ATLAS) (an improved version of ASSESS) is a software-based program used to compute the most vulnerable paths for both outsider adversary and violent insider attacks. The most vulnerable paths are computed in two different ways. The first minimizes P_I. This is called the CDP approach, because it is based on locating the critical detection point (CDP). The second minimizes delay after the practical detection point (PDP). These two analyses are complementary analyses approaches. The primary approach is the CDP approach. The PDP approach may identify paths that the CDP approach may not. A PDP analysis should never be performed without also performing comparable CDP analyses.

Another analysis feature identifies elements that are critical to the overall protection system effectiveness. Critical elements that, if individually degraded to a critical performance level on entry, will reduce the P_E.

---

[9]   A discussion of Markov mathematics is provided in SAND 2007-5591.

6.0   Vulnerability of Integrated Security Analysis (VISA) Method

The Vulnerability of Integrated Security Analysis (VISA) Method is a systems approach to a vulnerability analysis (VA). The VISA manual provides guidance for VA teams to perform tabletop VAs. A copy of the VISA manual may be obtained by contacting an NRC security specialist.

Appendix D

Blast Effects

APPENDIX D BLAST EFFECTS

A discussion on blast effects can be found in the NRC Regulatory Guides 5.68, "Protection Against Malevolent Use of Vehicles at Nuclear Power Plants" (SGI) and 5.69, "Guidance for the Application of Radiological Sabotage Design-Basis Threat in the Design, Development, and Implementation of a Physical Security Protection Program that Meets 10 CFR 73.55 Requirements" (SGI).

NRC FORM 335
(12-2010)
NRCMD 3.7

U.S. NUCLEAR REGULATORY COMMISSION

BIBLIOGRAPHIC DATA SHEET

(See instructions on the reverse)

1. REPORT NUMBER (Assigned by NRC, Add Vol., Supp., Rev., and Addendum Numbers, if any.) NUREG/CR-7145

2. TITLE AND SUBTITLE

Nuclear Power Plant Security Assessment Guide

3. DATE REPORT PUBLISHED	
MONTH	YEAR
April	2013

4. FIN OR GRANT NUMBER
NRC-42-07-036

5. AUTHOR(S)

J. Zamanali and C. Chwasz
D. Greenhalgh and J. Crockett

6. TYPE OF REPORT

Technical

7. PERIOD COVERED (Inclusive Dates)

8. PERFORMING ORGANIZATION - NAME AND ADDRESS (If NRC, provide Division, Office or Region, U. S. Nuclear Regulatory Commission, and mailing address; if contractor, provide name and mailing address.)

Nuclear Systems Analysis Operations Center

Information Systems Laboratories, Inc.,

11140 Rockville Pike, Rockville, MD 20852

9. SPONSORING ORGANIZATION - NAME AND ADDRESS (If NRC, type "Same as above"; if contractor, provide NRC Division, Office or Region, U. S. Nuclear Regulatory Commission, and mailing address.)

Division of Security Policy, Office of Nuclear Security and Incident Response, U.S. Nuclear Regulatory Commission
11545 Rockville Pike, Rockville, MD 20852-2738

10. SUPPLEMENTARY NOTES

11. ABSTRACT (200 words or less)

This document provides detailed guidance for the format and content of a security assessment of a commercial nuclear power plant.

The U.S. Nuclear Regulatory Commission (NRC) encourages design certification and combined license applicants to use this guidance to optimize physical security during the design phase. The expected result is a more robust security posture with less reliance on operational programs (human actions) and potentially costly retrofits. The NRC also encourages operating reactor licensees to use this guidance in planning and executing changes and upgrades of physical protection systems at existing sites.

12. KEY WORDS/DESCRIPTORS (List words or phrases that will assist researchers in locating the report.)

Physical Protection systems
Physical Security
Assessment
Design Certification

13. AVAILABILITY STATEMENT
unlimited

14. SECURITY CLASSIFICATION
(This Page)
unclassified
(This Report)
unclassified

15. NUMBER OF PAGES

16. PRICE

NUREG/CR-7145

Nuclear Power Plant Security Assessment Guide

April 2013

www.ingramcontent.com/pod-product-compliance
Lightning Source LLC
Chambersburg PA
CBHW081831170526
45167CB00007B/2782